Biology and the Social Crisis

BIOLOGY AND THE SOCIAL CRISIS

A social biology for everyman

J. K. BRIERLEY
D.Phil (Oxon), F.I.Biol.

With a Foreword by
C. D. DARLINGTON, F.R.S.

RUTHERFORD • MADISON • TEANECK
FAIRLEIGH DICKINSON UNIVERSITY PRESS

First American edition published 1970
by Associated University Presses, Inc.,
Cranbury, New Jersey 08512

Library of Congress Catalogue Card Number: 71-120071

Part title illustrations by Vernon Lord

ISBN: 0-8386-7719-3

Printed in the United States of America

To close members of my family tree
with love and respect.
Walter and Amy Brierley
Ann
and the children:
Sally,
Alistair
and
Christopher

Foreword

by Professor C. D. Darlington, F.R.S.
Fellow of Magdalen College, Sherardian Professor of Botany,
University of Oxford

PEOPLE who think about the affairs of mankind have always had plenty of troubles to occupy their thoughts. War, famine and pestilence have never been far from sight. The errors of the rich, the sufferings of the poor, and certain prospects of decay have always been evident and deplorable. But at the present time, the things that trouble us are of a new kind. History has taken a new turn. Our difficulties are of a new kind.

What upsets us now is people. There are beginning to be too many people in the world. And it is not simply a matter of numbers. These people are of many kinds. They are of very different kinds. And they don't all love one another. They don't even all understand one another. Far from it. Yet they are all *involved* in one another. That was clear to John Donne a long time ago and it is a hundred times clearer today.

Now if we ourselves are all tied up with one another, our problems are no less so. Which means the problems of numbers, of differences, and above all the problem of getting on together. They are part of the one great affair for all of us of birth, life and breeding.

The time has come, therefore, for those of us who are willing to think about these questions to turn our thoughts to some purpose. When we do so we discover that these questions are not new questions at all. How to increase the numbers of men or how to diminish them; how to prevent inbreeding or avoid outbreeding; how thereby to fix or to change the relations of races and classes; these are matters on which kings and priests have delivered their judgments and issued their decrees since the beginning of history. Over a period of four thousand years their laws have been recorded for us. Manu, Moses and Mahomet in Asia, and, in Europe, Roman Emperors from Augustus to Justinian and Roman pontiffs down to the present day have told us how we ought to breed and why we ought to multiply.

Throughout history these rules of breeding have provided the solid meat of moral teaching and religious belief. Indeed concern with breeding—or, as it is now called, SEX—has often seemed to be an obsession. But if we follow this concern back to an earlier stage of our evolution we find something different. Before agriculture had given man a regular supply of food he had quite another set of customs, morals and beliefs. Then already he was concerned with breeding. But then he was concerned to avoid its effects. His interest was in birth control, abortion and infanticide. He was indeed concerned more with restricting numbers than with increasing them. Just as we are.

Our problem is therefore a very old one. It has happened before. It is new merely in its urgency. It is now a crisis due to the sudden reversal of our needs. Our ancestors could scarcely have known what to do in such a crisis. Fortunately we are beginning to be equipped to deal with it. What we have learnt, beginning with Darwin and Mendel, is a method of thinking, a new method, which we need to help us in our emergency.

This is where Dr Brierley comes in. In this book he tells us what we need to know and how we need to think if we are to do anything about our emergency. It is the bare bones of the story. He shows us the dangers. He reveals the opportunities. Exciting or alarming we may find them both. Health and disease, education and crime, heredity and selection all appear in their connections. So many people have talked about them separately. But here they are where they belong, parts of one picture.

Dr Brierley gives us his opinions on all these matters. But he does not tell us what our conclusions should be. Rightly so. For all these thousand years or hundreds of generations our forebears have been working on how to reform the laws of breeding to meet the needs of civilization. We have less time. We can scarcely allow ourselves one generation to accomplish our reform or revolution. How are we to do it? Let us not be too hasty, Dr Brierley says, in giving our answer. But he implies, let us not waste time in putting the question. And he shows us how to do it.

Acknowledgments

THIS is not an original book. It is based on the work of many people. Any originality lies in the choice of material. Its ideas are derived mainly from three giants of the last century—Darwin, Galton and Mendel; and four of this—C. D. Darlington, J. B. S. Haldane, R. A. Fisher and J. S. Huxley. I owe a special debt to Professor Darlington not only because he has on occasions spent time discussing with me the wider issues of Man's biology but because he has read the manuscript and written the Foreword. But more than this, his own writings and speculations have seemed to me always to shed a fresh and penetrating light on many of the social problems of today both large and small.

I am grateful to Dr V. S. Butt, Fellow of Pembroke College, Oxford, Professor Philip Sheppard, F.R.S., Dr Geoffrey Gray of the Institute of Experimental Psychology, Oxford, Mr F. C. Minns, my former colleague at Manchester Grammar School and my colleague Mr J. E. H. Blackie, C.B., H.M.I., for reading parts of the draft manuscript and offering advice and criticism. I owe a great debt to my wife and my father who have read the manuscript for clarity of style and sense and have left it much improved. None of the above, however, can be responsible for any of the views expressed nor for any errors that may be found. I am grateful, too, to the numerous authors of the charts and tables acknowledged in the text and particularly Mr C. G. Vosa of the Botany School, Oxford, and Dr M. D. Casey of the Centre for Human Genetics, Sheffield, for allowing me to reproduce some of their photographs of human chromosomes.

JOHN BRIERLEY

Acknowledgments

Contents

Acknowledgments *page* ix
Prologue xv

PART 1: Human Heredity

1. How heredity works (1) Some fundamentals 3
2. How genes work 20
3. How heredity works (2) Male and female 25
4. Mutation 33
5. The origins of human individuality 38
6. Chromosome mistakes 47
7. The inheritance of genius 49
8. Marriage and close marriage 55
9. Natural selection in man 64
10. Eugenics 75

PART 2: Race

11. Race (1) Pinks, blacks, browns and yellows 83
12. Race (2) Biological engineering 93

PART 3: Man's Health and Food

13. Diseases of the affluent society 103
14. Affluence and neurosis 112
15. The iceberg of disease 117
16. Diseases of poor countries 123

17. What people eat *page* 130
18. Some myths about food and health 135
19. Horizon medicine 140

PART 4: The Crisis of Numbers

20. Population problems 153
21. Population control 167
22. Destruction of habitat 174
23. City life 186

PART 5: Youth and Age

24. Crime 197
25. Gangs 203
26. The Spermatozoon and the Spirochaete 208
27. Old age 215

PART 6: Brain and Behaviour

28. The Physics and Chemistry of the Brain 223
29. Doors in the Wall 233
30. Man's Behaviour 238

Bibliography 252
Index 257

List of Plates

facing page

1. The mitotic complement of a normal human female showing the 46 chromosomes 60

2. Hapsburg family 61

3*a*. Cell from a blood culture of a male with an extra Y chromosome.
b. The same *XYY* cell as in 3*a* but with the chromosomes arranged according to the Denver classification 76

4. The mitotic complement of a mongoloid human female showing 47 chromosomes 77

A million million spermatozoa,
All of them alive;
Out of their cataclysm but one poor Noah
Dare hope to survive

And among that billion minus one
Might have chanced to be
Shakespeare, another Newton, a new Donne—
But the One was Me.

ALDOUS HUXLEY

Mankind . . . will not willingly admit that its destiny can be revealed by the breeding of flies or the counting of chiasmata.

C. D. DARLINGTON
Royal Society Tercentenary Lecture 1960

Prologue

'THE year 2000 should find mankind well on the way towards exploitation of the solar system. This is the challenge of our century. We must not limit our vision, suppress our curiosity or dilute our determination.' So writes one of the world's greatest experts on space travel.* But is he right? And what is more important, are his priorities correct?

For man to exploit planets without excessive difficulty they will have to be very like the Earth in every way, so that man will be able to live and work on them comfortably and enjoyably. This means having an atmosphere, temperature range, light intensity and gravity pull, and many other things that man's body and the crops he grows are adapted to. There would need to be night and day, seasons, a sun of about the same size and appearance, green plants to provide the mainstay of the animal world as well as a source of free oxygen. The creature which must be absent is one as intelligent as man. If such an occupant of a planet existed conflict would certainly arise.

Readers who wish to know more about habitable planets for man should read the book by Dole.† The author gives a list of 14 stars in the immediate vicinity of the sun most likely to have habitable planets in orbit about them. It seems highly probable that there is at least one which would fulfil the conditions necessary for life. And this is only a small part of the universe. We often gaze up into the night sky and wonder whether there is life there. There is not a shred of direct evidence on this important question just as there would be none about our planet or its sun from a distant star. From Alpha Centauri 4·3 light years away an acute observer would see the same star pattern as we see from the Earth but there would be one large bright star added to

* Werner von Braun (see Bibliography). † See bibliography.

the zigzag of Cassiopeia. It would be our sun. There would be no evidence that it had in orbit around it a small outlying planet supporting a large brained upright mammal with unspecialized limbs whose high intelligence was beginning to turn towards the exploration of space.

Whether there are habitable planets or not, by 1970 man will have his foot on the first rung of the space ladder. By then our first astronauts will be stepping out on to the dead world of the moon to find, no doubt, a hopeless, airless, desert, alternatively torrid in the sun and bitter from its lack. By the year 2000 it is very likely that scientific colonies will be established on the moon. And perhaps in a million years man will have explored the entire Milky Way (our own galaxy), colonized any habitable planets and evolved away from the parent stock.

Let us hope he achieves these objects if he wants to, but before this he will need to set his own house in order. This book describes some facts about the biology of man in the middle of the twentieth century. The reader is intended to be left with the knowledge that there is plenty to be done on this planet. The most urgent problem is that of population, for unchecked fertility is at the root of most of the social, economic, and health problems the world is facing. Over-population has caused the growth of huge cities with their traffic congestion, special health hazards, and all the tensions of urban life; it has caused war and political instability throughout man's history; it has been responsible for the deaths of thousands of babies and infants through hunger and malnutrition; since the dawn of civilization it has caused spoilation of soil followed inevitably by destruction of society. Each year we witness the decline of wild flowers and insects by spraying and read about the destruction of the habitats of mammals and birds throughout the world—a direct result of population pressure. In Britain our country lanes are becoming a dumping ground for worn-out cars. Many of the symptoms of over-population can be measured, but the load of day-to-day misery man bears through not having enough to eat or by seeing his children die needlessly in a stifling hovel, is incalculable.

Perhaps the most evil and long-term effect of over-crowding is its sapping of moral energy. Even now most of us feel very little guilt about famine in far-distant countries; indeed other people's troubles are quickly shrugged off. It is a human characteristic. Over-population may breed callousness and stifle compassion. People may be seen as a plague. Even the birth of a child may cause sorrow instead of joy, while the natural instinct to save life may be of doubtful value. Hunger, anxiety and a sense of hopelessness may gradually debase the value of the individual. Indeed, in time, natural selection may favour the individual without all the virtues on which civilization is founded—compassion, love, respect for life and for the individual, and set at a premium greed, avarice, indifference and cruelty. Mass extermination by war will become more likely for we know that crowding causes aggression in animals and ugly feelings in ourselves.

The way to halt this snowballing disaster is not to ship off surplus population to other planets as has been seriously suggested. Even if habitable planets existed it would mean, from this moment, shooting up into space about 150,000 people a day to keep the population numbers at their present level. If the operation were to start a hundred years from now 900,000 people would have to wave goodbye to the Earth each day to keep the world population marking time at 18 billions. Although the population problem is serious the picture is relieved by some shafts of enlightenment. Governments and other national agencies are recognizing the need to check fertility because by doing so economic and social goals will be reached sooner. But to get through to the individual and family level, which is what matters in the long run, is a slow job. It involves cutting through a tangle of marriage practices, religious beliefs, sexual behaviour patterns and cultural traditions; and it is doubtful whether science alone can clear away these brambles and briers. It can present the facts through education so that the individual can make the choice. But do we need to be more radical and ruthless so that the brake can be put down hard? Should we, as Sir Julian Huxley has urged, hand out contraceptives with food, or if a country asks for help with industrial plant make part of the bargain acceptance of advice and medical help on population control? One thing is certain: it is important to look at population growth and food production as an *ecological* problem, as a balance between total world resources and people, and not as a race between them. It is a disgrace that at present one half of the world is busily trying to get its weight down while the other barely scrapes together one decent meal a day.

Seven problems, the first two basic, the rest perhaps trimmings which follow, must be tackled soon and should have priority if by the year 2000 man is to live more profitably and happily on earth. Some of these are world problems, others are Western ones, but in time will be world-wide. They are, in the main, the concern of this book.

1. Soaring *population* must be controlled. Unless it is all other efforts to improve the lot of mankind will be wasted.

2. For the poorer half of the world, providing enough for them to eat and teaching the young to read as a first step are vital issues.
3. The *uniqueness* of the individual from the moment of conception needs to be recognized by those in power, so that human individuality does not go to waste. This means that better ways of discovering latent talent, artistic as well as scientific and mechanical, need to be devised and applied. At present, school selection tests are too often concerned with a single value scale based on I.Q. Clearly selection is necessary, *but this should be 'refined selection with re-selection'**—the facility to retest and correct errors right up the age scale. Hand in hand with this is the need for equal opportunity to enter a maximum variety of educational systems. As well as this both skill and foresight are necessary to guide young people into work that suits their capacities and personalities. The late J. B. S. Haldane underlined the fact that the success or failure of a work-oriented society may depend on 'the choice of men for the jobs and jobs for the men'.
4. If we wish to control and annul or direct our aggressive tendencies then aggression *between man and man* and in other animals needs close study. Why is it, for example, that men are too inhibited to kill with their hands yet they will drop incendiary bombs on sleeping cities? Why do some of us feel angry when we are overtaken by another car? To understand these problems means switching more effort from the exploitation of outer space to a study of the 'inner space' of our own mind and to applying the methods and language of students of animal behaviour to a comparative study of aggression over a broad front. Animal studies on the collective aggression of one community against another of the same species can represent, as Lorenz states, 'a model in which we can see some of the dangers threatening ourselves'.
5. *Traffic*, particularly the motor car, needs to be kept in hand so that it does not destroy our cities, destruction not only from the rumble and vibration of thousands of vehicles but by the more sinister erosion of human standards. The sheer indifference of people to derelict cars parked by the roadside or in open parking lots, and other sights of aesthetic squalor, encourages our increasing disrespect for beauty and history. The motor car is with us to stay. Consequently our cities need to adapt themselves rapidly to the traffic loads to come so that the motor car does not dominate the

* Professor J. M. Thoday, F.R.S.

visual scene entirely. The equipment for traffic—good motorways, the best cars, parking facilities—could be visually exciting; but equally could be an unattractive mess leading to the sterilization of land on an inefficient scale. No amount of planning will remove the threat of carcinogens in the air and stress diseases resulting from too much noise and danger.

6. The *diversity* of man's wild habitat must be conserved as must his mineral and other resources. Agricultural methods tend to replace diversity with simple crop systems and in the past this has had serious social and economic effects. Diversity according to Charles Elton is a buffer against the destruction and replacement of native species by plant and animal interlopers often brought in by accident from abroad. Such invaders find it hard to secure a niche in the complex webs of plant and animal life fostered by diversity. Let us therefore keep our lanes, hedges, mowing fields, heaths and copses along with our crop systems for beauty and pleasure *and* as an insurance policy against aggressive aliens.
7. Society has not yet adjusted itself to the increase in the proportion of *old people* brought about by better standards of living and medical care. Western societies tend to make people feel old and useless by retiring and then neglecting them. By doing so they hasten mental deterioration, for body and mind work closely together. Perhaps affluent societies have something to learn from certain primitive peoples in this respect for the latter provide the social conditions in which to use the experience and wisdom of old people for the benefit of the community.

Part 1: Human Heredity

A knowledge of heredity, the tendency of like to beget like, is basic to the understanding of man and society. Genetics, which is the organized study of both variation and heredity, makes sense, as we shall see, of why a man should possess a character in common with his maternal or paternal grandfather but which his own parents do not possess. It helps to explain the causes of similarities and differences between himself and his brothers and sisters, not only in such obvious characters as eye colour but in blood groups, body build, fertility, intelligence, temperament and behaviour. It helps to explode the myth that if a mother sees something frightening during pregnancy her baby will be affected. Fortunately heredity is 'harder' than this. Genetics poses some important questions for the individual and society: how much does the minute submicroscopic world of genes and enzymes influence the whole man, the thinking, behaving 'self'? How much can the effects of the environment blur the imprint of heredity? Do people unconsciously choose the environments (the society they live in, the company they keep, the kind of work they do) that best suit their heredities?

Human genetics is still at the foundation and scaffolding stage. The solid foundations were laid by scientists working on peas, flies, wheat, beans, grasshoppers and moths—not on man. For example Mendel's work on peas gave us a simple understanding of human inheritance; the study of shades of colour in wheat provided the clue to the finely graded human characters like fertility, intelligence and height; the puzzle of maleness and femaleness was not sorted out until the chromosomes of grasshoppers and plant bugs had been looked at. Much patient observation over the years on mundane organisms such as these and others has put man in a position of beginning to know more about himself but it is important to recognize that there is much to be discovered and the scaffolding may need to be reconstructed as new truths are discovered and perhaps old ones demolished. Whether he will ever use this knowledge to prevent genetic decay or promote genetic improvement is a matter for the individual and society.

X
Y

1: How Heredity Works
(1) Some Fundamentals

LOOK around your family and see how much you are like them and how much you differ. You might have grandfather's eyes and temper, your sister might be the image of her mother but your brother seems to be the odd man out and like nobody you know. A century ago no one, except Gregor Mendel, an Austrian monk, had an inkling of how to explain, on a scientific basis, the causes of these likenesses and differences. Mendel's work on heredity was done on peas. We know now that what applies to peas, in the laws of inheritance, applies also to man, moulds, maize and any other living thing. Mendel published his work on heredity in 1865. Like most great discoveries it met with a chilly reception from the conservative scientists, and was forgotten. In 1900 when Mendel was dead, his work on heredity was rediscovered and its full significance realized.

CHROMOSOMES

Before some of the facts about Mendelism are discussed, it might be useful to describe some structures which, though unknown to Mendel, entirely supported his theory. These are the chromosomes; minute coiled bodies found in double sets in the nucleus of every cell of the body of plant, man or animal. They can only be seen under a good microscope. In ourselves in each cell there are 23 pairs, 46 in all (see Plate 1); the members of each pair being alike in shape; one set comes from the mother, the other from the father. When a human or any other creature grows, its cells divide to make the growth, and when the cells divide so do the chromosomes so that every cell always has the double pack of chromosomes. A different type of chromosome division happens in the formation of the reproductive cells or gametes. The production of these cells involves a reduction in the number of chromosomes so that each gamete has half the original number, e.g. 23 in man. The chromosome sets separate so that each gamete gets only one pack. The particular assortment of chromosomes in each half pack is a matter of chance, but the number of alternatives is restricted by a special shuffling mechanism. This works by dealing *one* member of each chromosome pair into different gamets so that if we imagine the 46 letters below to represent the 23 pairs of chromosomes *A* might be dealt into one gamete and *a* into another. Simultaneously of course the other pairs of letters (chromosomes) will also be separating. Now it will be clear that for any one gamete (egg or sperm) to be like another will be a rare event. To illustrate by a simple case. Suppose we had only two pairs of chromosomes represented by:

$$\begin{matrix} A & B \\ a & b \end{matrix}$$

When gametes are formed, four different chromosome combinations could result: *AB*, *Ab*, *aB*, *ab*. This would give four *unique* eggs or sperms. Since in reality we have 23 pairs of chromosomes

$$\begin{matrix} A & B & C & D & E \\ a & b & c & d & e \end{matrix} \text{ etc., up to 46}$$

It will be appreciated that the number of possible permutations of chromosomes in any one half pack in each gamete will be very great. The probability of any one being like another is in fact 2^{23}, or 1 in 8,384,608.

It is plain that the tiny chromosomes have the lion's share in deciding whether we are to be ugly or beautiful, stupid or intelligent, since they are the only things that are contributed equally by both parents to new life. It is from the fusion of the two cells, the tiny sperm and the large egg, that our life starts and the seal is set on our uniqueness by the shuffled chromosomes. Once the double number of 46 chromosomes is restored in the fertilized egg the chromosomes divide as does the cell containing them. Two cells, each with a double set of 46 chromosomes are thus formed. These divide to form four cells and these to form eight and so on until millions of cells are formed. When a baby is born it has about 25 trillion cells, each of which bears that double unique pack of chromosomes.

MENDELISM AND CHROMOSOMES

A summary of the interpretation of Mendel's results is as follows:

1. Inherited characters are controlled by factors (now called genes) contained in the cells of an organism.
2. The factors are always in *pairs*, one of the pair having been obtained from the father and the other from the mother. The pair of genes controlling a particular character may be alike or unalike. Thus, for example, they may be *AA*, *Aa*, or *aa*. Sometimes one of the genes of a pair is *dominant* over the other (e.g. *A* over *a*). This means that when *A* is present its instructions are obeyed rather than those of *a*, when the two occur together *Aa*. Eye colour in humans is controlled in this way (see below). The suppressed gene *a* is called *recessive* since the character it controls does not appear. Where both the genes of the pair are alike, as in *AA* or *aa*, their effect on the individual is in no doubt. Sometimes instructions from both dominant and recessive genes are obeyed to give an intermediate effect.
3. In the formation of sperms and eggs the members of a gene pair part company or *segregate* and either one *or* the other, never both, is passed in to each gamete; which member of the pair is a matter of pure chance. Pairs of genes are restored again at the fusion of egg and sperm. In the body of an individual the pairs of genes do not blend, like ink and water, or get diluted or become contaminated with each other, for, in the formation of gametes they always segregate as pure as when they arrived at fertilization.

We now know that genes are borne on the chromosomes, arranged end to end in a definite order, and move along with the chromosomes. They are paired and, like the chromosomes, they separate in the formation of the gametes. As with the pairs of chromosomes, it is a matter of pure chance which member of a gene pair goes into a particular gamete.

SKIPPING A GENERATION
FAMILY TREE FOR EYE COLOUR

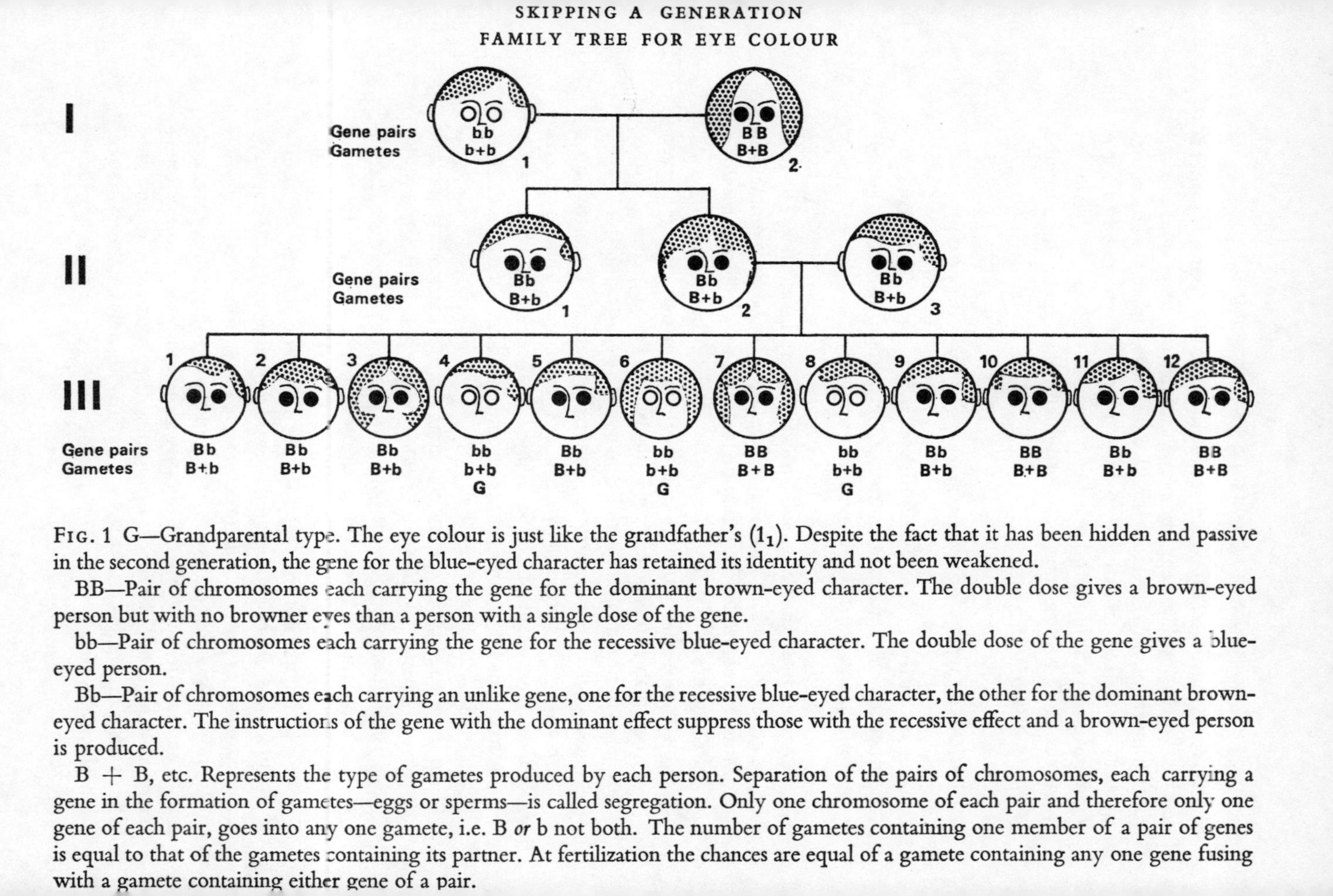

FIG. 1 G—Grandparental type. The eye colour is just like the grandfather's (1_1). Despite the fact that it has been hidden and passive in the second generation, the gene for the blue-eyed character has retained its identity and not been weakened.

BB—Pair of chromosomes each carrying the gene for the dominant brown-eyed character. The double dose gives a brown-eyed person but with no browner eyes than a person with a single dose of the gene.

bb—Pair of chromosomes each carrying the gene for the recessive blue-eyed character. The double dose of the gene gives a blue-eyed person.

Bb—Pair of chromosomes each carrying an unlike gene, one for the recessive blue-eyed character, the other for the dominant brown-eyed character. The instructions of the gene with the dominant effect suppress those with the recessive effect and a brown-eyed person is produced.

B + B, etc. Represents the type of gametes produced by each person. Separation of the pairs of chromosomes, each carrying a gene in the formation of gametes—eggs or sperms—is called segregation. Only one chromosome of each pair and therefore only one gene of each pair, goes into any one gamete, i.e. B *or* b not both. The number of gametes containing one member of a pair of genes is equal to that of the gametes containing its partner. At fertilization the chances are equal of a gamete containing any one gene fusing with a gamete containing either gene of a pair.

Let us take a simple hypothetical human example of blue and brown eye colour to illustrate the argument so far. This simple example is used with the knowledge that the inheritance of eye colour is, in fact, more complex than it appears. There are up to 100 kinds of eye colour and there is some suggestion that sex-linked inheritance may play a part (see Chapter 3); eye colour also changes with age.

Suppose that a man with blue eyes marries a woman with brown eyes and that all their children had brown eyes. Then let us suppose that one of these children, a brown-eyed man (it could just as well be a woman), marries a brown-eyed girl who is the child of a similar couple and they had twelve children. Of these, nine had brown eyes and three blue eyes like their grandfather's. How can we explain this family tree using our knowledge of the working of chromosomes and heredity? A glance at the tree (see Fig. 1) shows two interesting features. First that blue eyes are present in the grandfather's generation, disappear in the second and reappear in the third generation, among the grandchildren. Second, that in this generation the grandchildren have blue eyes and brown eyes in the ratio of 1 : 3—three blue-eyed children to nine brown-eyed ones. The family tree, then, illustrates an important point, a character *skipping a generation*, for the blue eyes present in the first generation vanish in the second and appear in the third. From this evidence we can infer that the blue-eyed character is controlled by a *recessive* gene whose instructions are overwhelmed by those of the *dominant* brown gene. But why do blue eyes re-appear and why the 1 : 3 ratio?

Let us suppose that the ability to produce brown eyes is due to the action of a gene which we can call *B*. It may exist in another form *b* which is recessive and which can only instruct a small amount of pigment to form in the iris, giving blue eyes. The gene pair is carried on a pair of similar chromosomes, that is *B* on one chromosome and *b* on another. As we have seen, the chromosomes separate from each other in egg and sperm formation, one member of each pair being dealt into each gamete. The separation of the chromosomes along with the genes they carry brings about segregation so that every gamete contains one member of a pair. The brown-eyed father of the first generation carries a double dose of the *B* gene (*BB*) and produces sperms containing a single *B* gene while the blue-eyed grandmother carries a double dose of the recessive *b* gene (*bb*) and produces eggs which contain *b* only. Fusion of eggs and sperms will restore the pairs of chromosomes and genes to produce all *Bb* children and these will all be brown-eyed yet will be carriers of the hidden recessive gene for blue eyes. If two such people marry, as in the second generation of our family tree, the sperms and eggs produced by the man and woman will be *B* or *b* type. Millions of sperms are produced by the man, with *B* and *b* types in equality. This must be so, for as we have seen the automatic cell machinery in gamete formation is arranged so that each member of a pair of chromosomes, and the genes they carry, is dealt into different gametes. The dividing machinery keeps going so that two or two million sperms would be half *B* and half *b* type. The same mechanism works in egg formation. The children of this marriage, our third generation, will throw up the blue-eyed character for this reason. Namely, that the chances are equal that a sperm carrying *B* will meet an egg carrying *B or b*, so producing fertilized eggs (zygotes) of the type *BB* and *Bb* in equal numbers. Similarly with the equally numerous sperms carrying *b*, when these meet the two types of egg, zygotes of *bb* and *Bb* are produced also in equal numbers. We then get three types of fertilized egg (and eventually people); *BB*, *Bb*, *bb* in a ratio of 1 : 2 : 1. These are the proportions always produced from the *chance* combinations of two continually alternative types. If two pennies are tossed together a great many times, the throws give 25% both heads, 25% both tails and 50% one head and one tail, as in the genetic ratio. However, in the family tree the three classes of fertilized egg, *BB*, *Bb*, *bb*, will not give children with brown eyes, intermediate coloured eyes and blue eyes in a 1 : 2 : 1 ratio because the presence of a *B* gene even in a single dose causes brown eyes to form. Consequently, only two classes appear, brown and blue eyes in a 3 : 1 ratio, representing a 1 : 2 : 1 ratio with the first two ratios added together. This 3 : 1 ratio is, of course, only seen when large numbers are studied. One or two families lumped together, even if they are large, may not show it. With plants and animals, where large numbers can be bred for two generations, the ratios can be very clear.

This example clears up any puzzle over the skipping of a generation by some characters. The gene for a character like blue eyes or red hair that skips a generation is transmitted, but it is recessive, so that the character itself cannot re-appear until the second generation at the earliest. Moreover, the pedigree illustrates Mendel's most fundamental discovery, the foundation on which the whole of modern genetics is built. This is that the basis of heredity consists of material units in the sperms and eggs; that these units or genes are self-perpetuating and self-copying and do not themselves blend like ink and water when crossed. They pass down the stream of life unchanged and a characteristic which might not have appeared in a family for two or three generations, or even longer, may suddenly reappear.

MENDEL'S LAWS AND US

The evidence that Mendel's principles of heredity apply to man is clearly abundant if we look around us. The manifestation of recessives (the double dose of a recessive gene is necessary, you may remember, before the character appears) is often due to close marriages, as between cousins (see Chapter 8). For example, probably the first recorded case of albinism (a typical albino has white hair and pink eyes) is that of Noah (see Fig. 2) recorded in the Book of Enoch, written in the first and second centuries B.C. (illustrating, incidentally, the antiquity of the condition). The birth of Noah is recorded as that of a miraculous child but in fact the account is a vivid description of an albino:

'For to my son Lamech a child has been born, who resembles not him; and whose nature is not like the nature of man. His colour is whiter than snow; he is redder than the rose; the hair of his head is whiter than white wool; his eyes are like the rays of the sun.'

One account, in the Book of Jubilees, shows Noah as the child of Lamech and his first cousin Bt'nws. Lamech must have inherited his hidden recessive gene from his father Methuselah, while Bt'nws inherited it from one of her parents who is recorded as a brother or sister, it is not known which, of Methuselah. Bt'nws and Lamech not only produced Noah, the boat builder and sailor, but made genetic history.

Of course, the outcropping of such hidden recessives in a child is very rare. The human species carries many of these recessives under cover (see Table 1) but it is only when two carriers produce offspring that the character shows itself, on the average in a quarter of the children. Two-thirds of the normal children will carry the hidden recessive with no obvious effects. It is quite likely that the 'carrier' parents may never produce an abnormal child if they produce only a small number of children. On the other hand, the first-born could by chance be defective. Table 1 makes it clear that even though the frequency of the carriers of a 'bad' gene may be high the chances of producing an affected child (i.e. two carriers marrying) is surprisingly low.

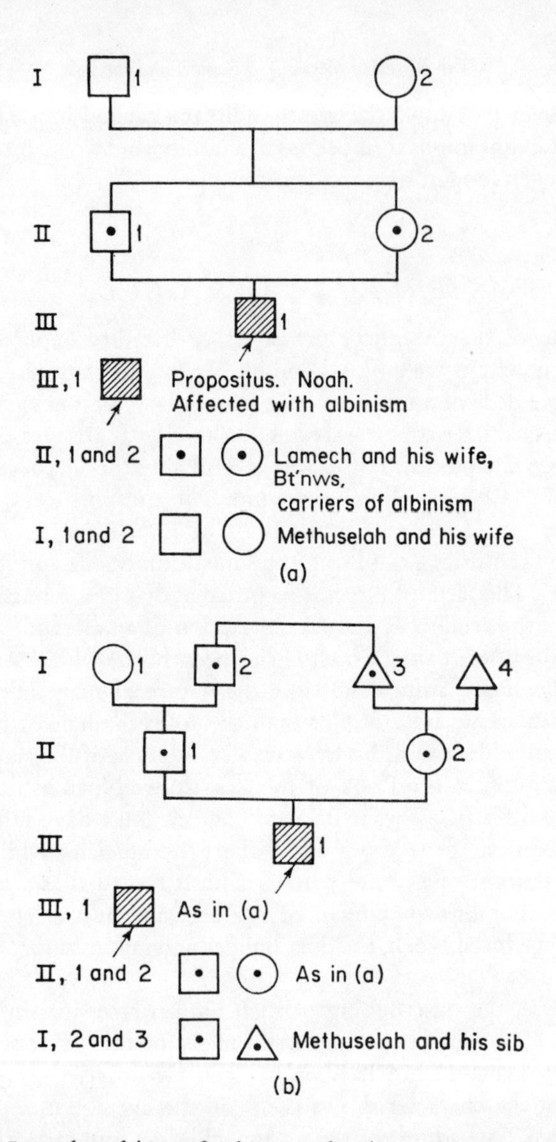

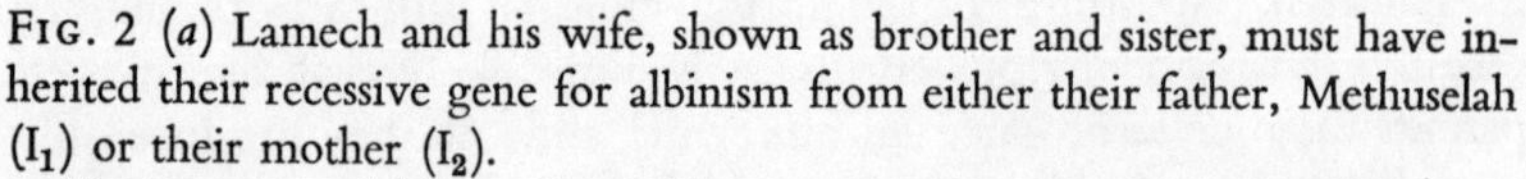

FIG. 2 (*a*) Lamech and his wife, shown as brother and sister, must have inherited their recessive gene for albinism from either their father, Methuselah (I_1) or their mother (I_2).

(*b*) Lamech and his wife are shown as first cousins. Lamech must have inherited his recessive gene from his father, Methuselah whilst Bt'nws inherited it from one of her parents, who is recorded as a sib of Methuselah.

After Sorsby, A., *Noah—an Albino*, *British Medical Journal*, 27 December 1958.

TABLE 1

Disease controlled by recessive gene	*Frequency of an affected person in population*	*Frequency of carrier in population*
Fibrocystic disease of pancreas (see p. 76)	About 1 in 2,000	About 1 in 25
Albinism (see p. 9)	About 1 in 20,000	About 1 in 72
Phenylketonuria (see p. 24)	About 1 in 25,000	About 1 in 80
Alkaptonuria (see p. 22)	About 1 in 1,000,000	About 1 in 502

On the 'bad' side, Mendel's principles apply to recessive genes controlling complete albinism, alkaptonuria, phenylketonuria and deaf-mutism. On the 'good' side there is recessive control of red hair, curly hair, and blue and grey eyes. These genes, like all recessives, can skip a generation—and are often said to 'run in families'.

We are on firmer ground with dominant genes because the trait is passed repeatedly from parent to child without the sinister skipping characteristic of recessiveness. If the trait is very disabling, there are no children. Looking around we often see dominant inheritance. For example, father and son, the one bald, the other with thinning hair; father and daughter with a forelock of white hair (see Fig. 3); a man and his son with piebald hair; a brunette, dark-eyed mother with a brood of dark children. It is obvious that these fairly common genes can do no harm. Nor do the rarer genes, giving an obvious family stamp like the jutting lip of the Hapsburgs (see Plate 2), borne by the descendants of the Emperor Maximilian, or the children of the Scipio Family, some with six fingers and others with normal hands. Scipio Africanus (and the Giant of Gath), it is said, had six fingers so the dominant gene has been handed down through centuries. Some genetic oddities are exhibited in circus shows. The thickset dwarf with short legs and the indiarubber man are rare examples of the effects of dominant genes. The frequency of the former condition appears to be about 1 in 40,000. A more common condition caused by a dominant

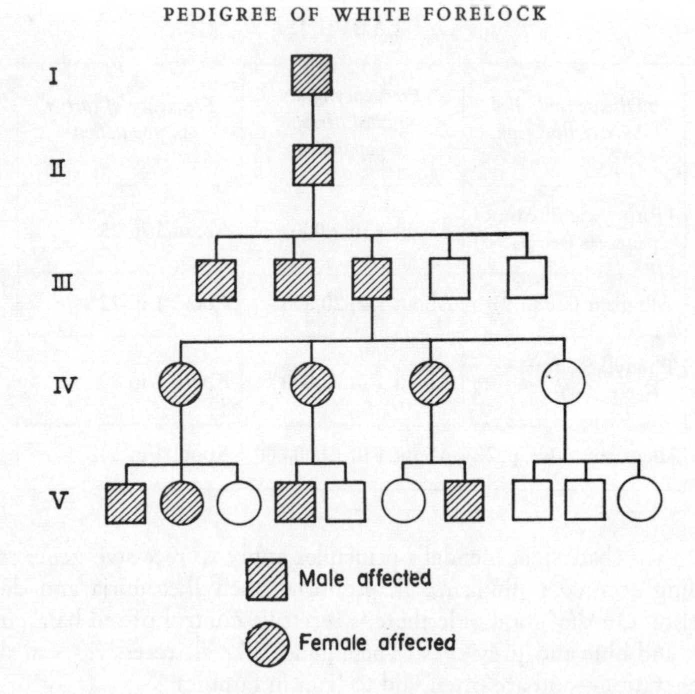

FIG. 3 Note that the trait is transmitted through five generations, that it is present in men and women and that it occurs in both sexes of the children of an affected parent. The pedigree omits symbols for the spouses of affected parents because in a rare trait it can be taken for granted that the spouses are not affected.

After Stern, C., *Principles of Human Genetics* (Freeman, 1960).

gene is that of gout which affects about 1 in 1,000. A curious illustration of a dominant trait is that of the 'porcupine man'. The gene causes the skin to become, in places, half an inch thick, black and very rough, while the hands, feet and face are normal. The first 'porcupine man' was born to the Lambert family in 1716 at Sapiston in Suffolk and was shown to a meeting of the Royal Society in 1734. Six of his family were affected, but five died in infancy. The remaining son had seven children and three sons were porcupine men. Two of these married but although a porcupine infant was born it died, the last of the line, in 1801. The men exhibited themselves in side shows and earned a good living.

Other dominant genes, happily very rare indeed, can cause misery and suffering. The 'lobster-claw' deformity, where hands or feet, or both are deformed in half the children of a carrier of the special gene, causes the hands or feet to become clawed with only two misshapen fingers or toes. A dominant 'time' gene, so called because it does not start working until between the ages of twenty-five and forty, is responsible for Huntington's chorea, a disease which causes repeated involuntary movements and eventually may produce insanity. In America the gene was carried in by immigrants from Bures in Suffolk who landed at Boston in 1630. In one case the line of descent could be traced unbroken through twelve generations to one of these immigrants. Some of the affected people were accused in the witch trials in Salem, New England, in 1692. As a result of the trials 19 people, mainly women, were condemned and executed for practising 'witchcraft' which may have been genetic madness.

SHUFFLING THE GENES

So far only one of Mendel's 'laws' relating to heredity has been dis cussed. A second great principle discovered by him is concerned with the inheritance of two or more pairs of genes which, when they segregate together on the chromosomes, the distribution of any one of them is independent of the distribution of the others. In other words, unless the two pairs of genes concerned lie on the same pair of chromosomes, they will segregate independently. For example, if a woolly-haired, dark-eyed man married a straight-haired girl with blue eyes all the children could be woolly-haired with dark eyes (see Fig. 4). The recessive characters, straight hair and blue eyes would disappear in the first generation. If a child of this marriage married a child of a similar couple, some of their children, if the family were large, say twenty, would be like the grandparents, but others would be novelties: brown eyes with straight hair and blue eyes with woolly hair—these are called *recombinations*. The gene pairs have shuffled; they have broken their old combinations and in some cases formed new ones.

The behaviour of the chromosomes, the carriers of the genes, when they separate into two packs in gamete formation, explains how the gene outfit of the parents gets automatically dismantled before reproduction. Although all the parental genes are transmitted their reappearance in the same combinations is very unlikely since, as

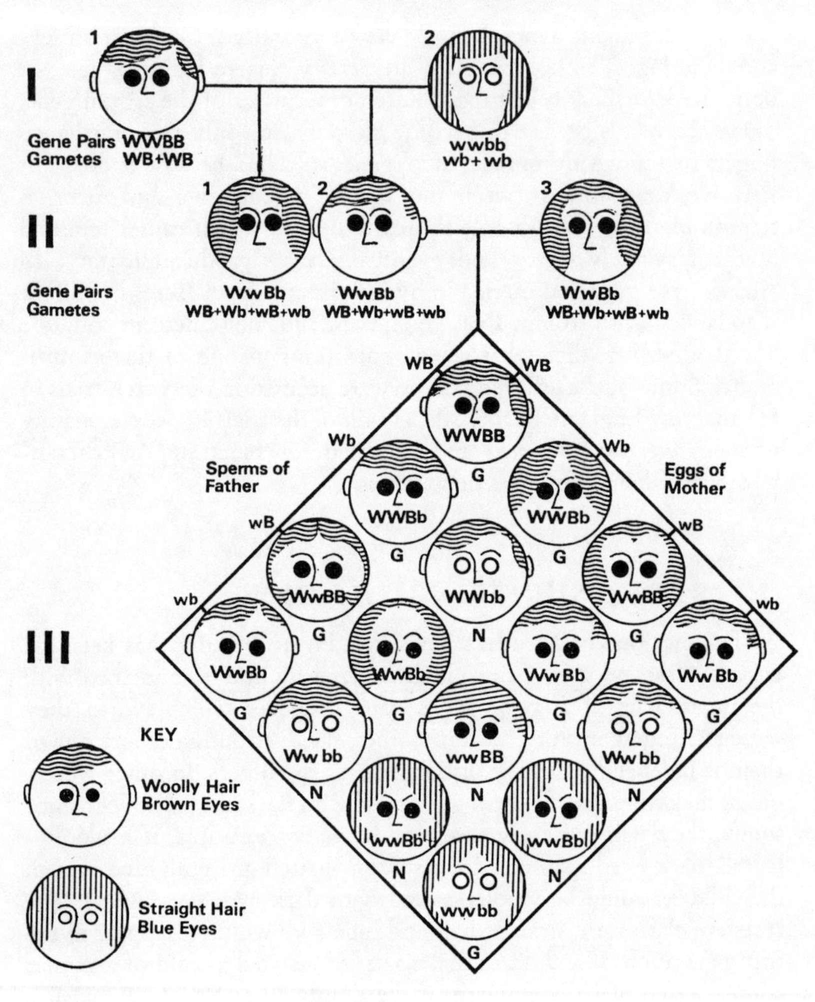

G—Grandparental types.
N—Novelties due to recombination of parental genes

FIG. 4 In this example two pairs of chromosomes and two pairs of genes are involved. The gene for woolly hair (W) is dominant over straight hair (w) and the gene for brown eyes (B) which is dominant over blue eyes (b). Thus WWBB represents two pairs of chromosomes which carry on one pair the genes for woolly hair and on the other the genes for brown eyes. When the pairs of chromosomes separate in gamete formation it is a matter of chance

previously described on page 4, one of each *pair* of chromosomes is dealt into two different gametes. It is purely a matter of *chance* which one of a particular pair goes with the other 22. This chance distribution of chromosomes into two half sets of 23 each makes it very unlikely that one gamete will be like another. The fusion of egg and sperm each containing a rearrangement of the parental gene outfits produces eventually a unique human being.

The example above, even with only two pairs of chromosomes and genes involved, shows some novel reassortments in eye colour and hair quality compared with the grandparental types. It is no wonder then, with 23 pairs of chromosomes whose gene outfits are automatically taken to pieces and reassembled before reproduction, that there are differences between brothers and sisters, that a genius may not have a genius as his son, that it is possible for 'ordinary' people to produce a high flier and scholarship winner and that a pair of ugly ducklings can produce a lovely child. In short, brothers and sisters are *genetically* alike or different depending on how many like genes they have and how many different.

---

whether for example W in II_2 or II_3 becomes included in a gamete with B or b, similarly for the equally abundant w genes. Thus four types of gamete are produced WB, Wb, wB and wb in equal numbers by each individual. These gamete types can be combined in 16 possible ways as the diagram shows. But owing to dominance they give rise to four distinct types only. Nine of the combinations contain at least one dominant member of both pairs of genes and so they have woolly hair and brown eyes, three have bb and at least one W and are therefore blue-eyed with woolly hair, three possess ww and at least one B so are straight-haired and brown-eyed, while one has a double dose of both recessive characters and is blue-eyed and straight-haired. This example illustrates the very important fact that from only two types (I_1 and I_2) in the original marriage four types of children are born in the third generation, two reversions to 'type' and two novelties. The family tree, and the theory behind it, helps to explain why brothers and sisters of any one family are different from each other. Remember that not only *two* pairs of chromosomes are segregating independently but 23 pairs. Many characters do not segregate as sharply as these described but manifest themselves as vaguer differences in families—height, 'intelligence' scores in mental tests, longevity, resistance to disease, blood pressure, temperament. These differences are controlled by polygenes which have small effects on a major character like height or fertility. These polygenes are lodged on the chromosomes in the same way as the 'Mendelian' genes and segregate in the same way, giving rise to the multitude of differences and similarities in families.

PLUS AND MINUS GENES

Many of the inherited characters described so far have been fairly clear-cut, like blue or brown eyes or straight or woolly hair or having or not having a particular disease or disability. These are all readily

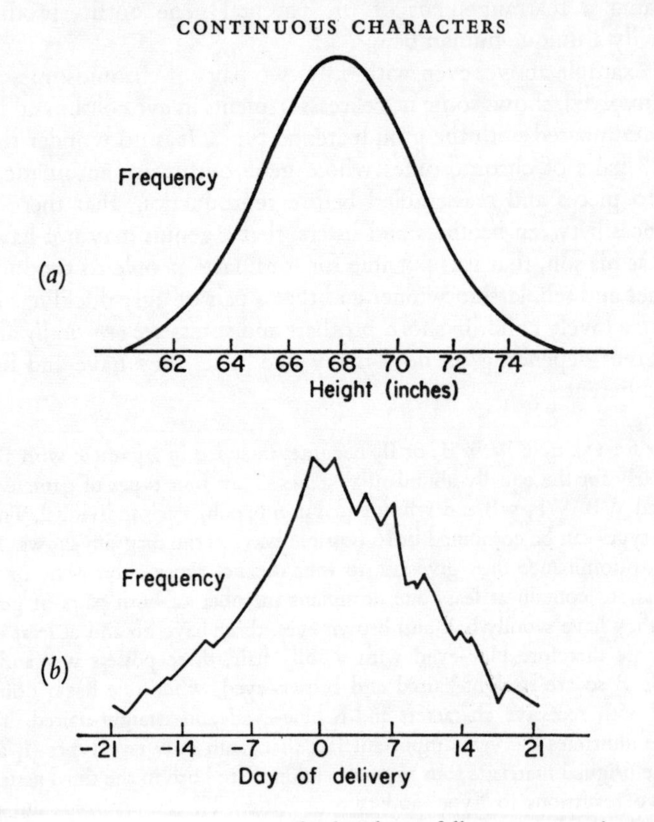

FIG. 5 (*a*) A distribution curve for height in full-grown men in south-east England.

After Carter, C. O., *Human Heredity* (Pelican Books, 1963).

(*b*) The graph shows there was a gradual increase in the number of confinements as the estimated date (O) approached. It is a measure of the length of pregnancy; it will be seen that most cases coincide with the doctor's prediction and that there are fewer and fewer cases in the more extreme ranges. The variation is due to polygenes and to environmental causes. The graph has not been 'smoothed' like the one on height but they both have similar shapes.

After Clarke, C. A., *Genetic for the Clinician* (Blackwell, 1962).

distinguishable. But there are much more important human character differences that are by no means clear cut, but run into each other. These are called *continuous* characters, like human height, fertility and intelligence. For example, a marriage between a tall man and a small woman does not produce clear-cut families of tall and short children because height is controlled by many genes producing small additive effects. A group of about 1,000 men, for example, will not fall into two or three sharply defined heights but will vary in stature over a fairly limited range and within this range very tall and very short men are uncommon. This effect is shown in Fig. 5(*a*) and the same sort of graph is obtained if the length of pregnancy of a large group of women is measured (Fig. 5(*b*)).

Francis Galton found the clue to the explanation of these rather indefinite kind of variations. He thought that finely graded characteristics like height or intelligence were controlled by a mosaic of small, inherited units. It has been shown, largely as a result of breeding work with small fruit flies called *Drosophila*, that these *polygenes* as they are called, are, like the obvious 'Mendelian' genes described, situated on the chromosomes but instead of producing a sharp, clear cut effect, they work in groups. Each member of a group has a small quantitative effect, plus or minus, on a major character like height. The graded curve for height shown above, is, or course, not only due to the work of polygenes affecting limb length and head and body sizes independently but to the effects of the environment; inherited tendencies plus environmental effects help to smooth the curve.

THE INEQUALITY OF MAN

What we see then when we look in a mirror, or at our children, is the result of the fashioning of heredity and of the environment before and after birth, whose work is so finely intermeshed that a seamless fabric results which makes man what he is and what he is to be. We see the results of the bolder work of dominant and recessive genes and the finer, more subtle touches of the polygenes. When a man speaks, the pitch of the voice may be like his father's or grandfather's and is inherited (see Fig. 6), but his dialect has been picked up from those around him and his views on any subject may be the fruit of human experience of centuries. These are obviously acquired through speech and books and are in man just as important as his genetic endowment.

How effectively a man communicates and learns, however, is regulated to a large extent by the inherited quality of his brain cells and even these can be weakened by an environment effect such as oxygen

'VOICEPRINTS' OF INDIVIDUALS

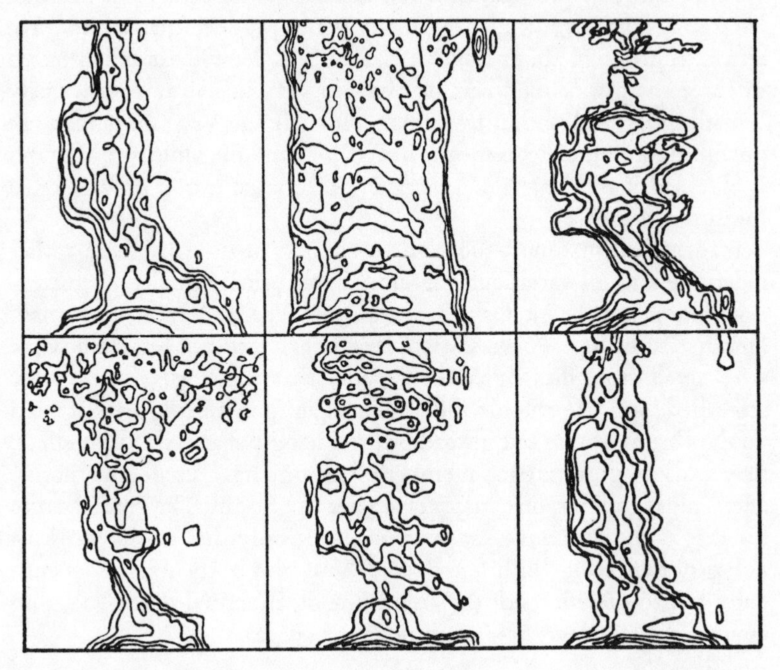

FIG. 6 These six 'voiceprints' are of five different people pronouncing the word 'you'. One person has said it twice.

The natural shape and size of a person's speech apparatus causes his voice energy to be concentrated into bands of frequencies. The pattern—but not the exact configuration—of these bands remains essentially the same even when a person attempts to disguise his voice by lowering or raising the pitch, speaking in a whisper, muffling the voice, or affecting an accent.

Time is plotted from left to right. That is, the beginning of the word 'you' is at the left, and the end of the word is at the right in each voiceprint. The lower pitch of sound appears at bottom and the higher pitch toward the top of a print. Greater intensity of sound at each frequency for a particular time is represented by peaks.

Despite his attempts to disguise the voice, the strikingly similar contours of the prints at upper-left and lower-right reveal it's the same man talking.

From *The Observer*, 17 April 1966.

starvation before birth. This close meshing of different natures and nurtures results in the unique nature of human beings. There are some scientists who believe in the overriding importance of heredity and some who think that environment is all important in making a man what he is. But to some extent this is a sterile controversy. The genes of a Milton or Darwin or a Bach would not have reached their ceiling had they been born in some village or town away from opportunity and stimulating company. Nor would a dullard change to a Newton even with access to the best libraries and endless opportunity. His inherited potential is low. The American Declaration of Independence stated that 'all men are created equal'. We have seen that there is very little evidence of this. Man comes in great variety and it is up to man himself to create a rich and varied environment that will allow the proper growth of varied inborn talents.

2: How Genes Work

GENES have been mentioned in the last chapter and it might be useful now to put forward some ideas about their fine structure and how they are thought to work. Let it be stated at once that genes, like electrons, have never actually been seen. They are theoretical structures which help us to explain heredity. It has been said that they are born on the chromosomes and, like the chromosomes, are paired and arranged in thousands, end to end in a definite order down the length of a chromosome. It has been shown too that the characteristics, of the hereditary units or genes which Mendel discovered are caused by the visible gymastics of the chromosomes (as seen under the microscope) in the special cell division before the formation of sperms and eggs and at fertilization. The genes, like the chromosomes that carry them—in fact we can think of chromosomes as simply rods of genes—are self-perpetuating and self-copying, for when a cell divides the two resultant cells and their chromosomes and genes are *identical*. Mendel showed that the genes never blend—although the characters they control might blend—nor do they wear out like the fading ink of an amateur's printing press, to make weaker and weaker copies of themselves. They separate from each other at reproduction as if freshly printed (to continue the printing simile). We have seen this when a character skips a generation and reappears in the next unchanged. None of this tells us much about how genes work or what they look like or how big they are. Some scientists have compared them with strings of beads, each bead being a single gene but modern research has shown that these big genes are really divided into smaller units each of which is a gene.

Professor Penrose has given us a better comparison than a string of bead-like genes. He suggests that we should take as our model a tape measure. The old notions of a gene would be a space of one inch on the measure but now the same distance would be found to contain tens or hundreds of genes. The smallest gene might be small enough to correspond to a cross-slice of a chromosome a millionth of a millimetre thick. Recently chemical and X-ray analysis has shown that a nucleic acid, D.N.A. for short, is found in the chromosomes inside the cell nucleus while another nucleic acid, R.N.A., is found in the cytoplasm, and can form proteins. D.N.A. chains or molecules can reproduce themselves exactly and in this way the genes are thought to reproduce, when chromosomes divide to form new cells. D.N.A. then carries the genetical programme of the cell and is thought to be the *architect* of enzymes and other proteins while R.N.A. is thought to be their builder. R.N.A. works in the cell cytoplasm from the D.N.A. blueprint of the chromosomes present in the nucleus. Proteins are the very stuff of life and form the bulk of blood, muscle, brain and the living parts of bone, while enzymes, which are proteins, control the rate of chemical processes in living cells. D.N.A. is thought to produce tailor-made plans for specific proteins and these are made into enzymes in the cell cytoplasm by R.N.A. working in conjuction with other special protein formers. Many enzymes are specific in that they do only one job so as not to interfere with other cell enzymes doing other work. It is absolutely necessary for D.N.A. to produce plans for different kinds of protein. Probably a particular length of D.N.A. (a gene) acts as the architect for a particular enzyme. But there are still many things not yet understood about the chemistry of the chromosomes. For example, how is the chromosome arranged into specific parts (genes), each of which performs specific jobs?

GARROD AND BIOCHEMICAL GENETICS

At one time it was thought that *one* gene produced *one* enzyme. To understand this we must go back more than 60 years to 1902, when Archibald Garrod drew attention to the hereditary disease of alkaptonuria. This is a rare disease where a substance, alkapton, accumulates in the urine, turning it black on being exposed to air. In children there are no symptoms other than the urine turning black but later in life the teeth and ear cartilages blacken and joint inflammation, due to bone degeneration, sets in. This condition is due to a recessive gene that must be inherited from both parents if it is to take effect. People with one normal dominant gene and one hidden recessive (*Aa*) would not show any disease but the children of two such people would include some *aa* types and the effect would be shown (see Fig. 7). Garrod suggested that affected people lacked the enzyme responsible for breaking down alkapton to carbon dioxide and water.

Garrod's idea was revolutionary. It fostered the modern development of biochemical genetics which uses as its experimental tools, yeasts and moulds. These organisms are bombarded with X-rays which

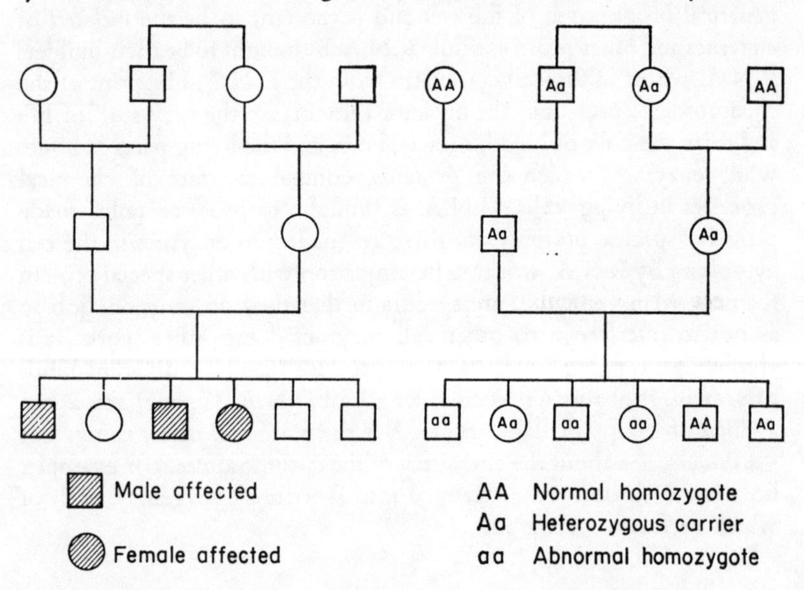

FIG. 7 Pedigree of alkaptonuria and its genetical interpretation
After Penrose, L., S., *Outline of Human Genetics* (Heinemann, 1963).

can cause alterations in otherwise stable genes. The alterations are called mutations and are the subject of Chapter 4. A normal mould (unlike an animal) builds all the amino acids it needs from simple materials. Some mutations have the effect of causing an originally normal mould or yeast to demand, for example, a particular amino acid (the building block of proteins) in their diet in order to grow properly. Such an experiment shows that the normal gene took part in producing an enzyme needed to build the necessary amino acid which the abnormal (mutant) gene failed to produce. It is likely that a chain of enzymes, each producing an amino acid, is necessary to produce a given enzyme protein. The alteration of one gene might cause, in some cases, a block in the reaction and the pile-up of an intermediate substance like alkapton that spills over into the urine. It is too simple to think of one gene producing one final product in total isolation from other genes. The whole packet of genes in the nucleus is necessary for the proper functioning of any one gene and the proper functions of the cell. It seems likely that all the genes together help to produce each enzyme but there is a special gene that gives the finishing touch to each product. This last touch makes an enzyme specific, that is able to carry out a particular reaction.

The hairs-breadth scale on which genes work (and on which the life of populations can depend) can be illustrated by the difference between normal haemoglobin–which is the red, iron-containing pigment essential for carrying oxygen round the body—and the sickle-cell type referred to in Chapter 9. The difference between this pair of haemoglobins, which is determined by a pair of alternative genes, is a single amino-acid. The normal gene instructs glutamic acid to be added to the haemoglobin molecular chain while the sickle-cell orders another amino-acid, valine to be substituted. Some 30 different haemoglobin variants are now known, some of which may be determined in this way. The important point is that here we have a gene which picks out precisely one amino-acid which, in the case of valine, is the means of survival of millions of people in malarious regions of the world. And just as remarkable is the fact that a double dose of the gene which orders valine to be made, kills.

GENES AND BACKWARD BABIES

When my small son was three months old a nurse called at the house for a routine talk about the baby's health; she also tested his wet nappy with a chemical indicator which told her if he was passing a special sugar in his urine. Luckily he was not but occasionally babies lack an essential enzyme for breaking down milk sugar (galactose) and the disorder is inherited as a recessive trait from both parents. Here again we have a gene–enzyme effect. Galactosaemia, as it is called, is a very serious baby disease which causes, among other things, mental backwardness. If galactose is removed from the baby's food it gets better but it must be done quickly or the brain is permanently injured.

Another rare disease caused by a recessive gene inherited from both parents is phenylketonuria and is due to the cutting out of the supply of an essential enzyme shortly after birth. The enzyme is necessary for the proper digestion of an essential food component, phenylalanine. The normal pair of dominant genes can make the enzyme—even one gene can make sufficient—whereas the pair of recessives cannot make the enzyme at all and the disease develops, causing severe mental backwardness and other effects such as very fair skin and a lighter hair and eye colour than other normal members of a family. Foods not containing phenylalanine are successful in arresting the disease which can be detected by a special chemical placed on wet nappies. Discoveries like these revolutionized ideas about the causes of mental deficiency (phenylketonuria is probably responsible for about 0·6% of the mentally defective populations of the world who are in institutions). Previously doctors had thought that backwardness was due simply to faulty brain development and not, as it is in these cases, due to the failure to produce an enzyme necessary for a particular chemical job. It is salutary to remind ourselves that much of what passes for normal behaviour and conduct depends perhaps on having the right genes and the right enzymes to do the necessary chemical jobs of the brain and body.

3: How Heredity Works (2) Male and Female

WHAT are the chances of having a boy next? I've got two girls already.' This is a common and important question to mothers and sometimes to kings, and a simple knowledge of genetics will help to explain it. Again some mothers think that if the baby is carried 'well forward' or if her own skin is mottled, the baby is bound to be a girl. If the baby lies to the back and the mother's complexion is pink and fresh many people think that 'a boy is on the way'. A simple knowledge of genetics throws a penetrating searchlight on these superstitions, to reveal a beautifully simple, automatic mechanism, quite out of our own control, that ensures a nearly equal supply of baby boys and girls, perhaps not in one family, but when the averages are taken across many families. We must go back to the chromosome story to explain the inheritance of sex.

Every living cell of the body, as was previously described, contains 23 pairs of chromosomes. In a baby girl and a woman, the members of one of these pairs are called the *X* chromosomes and lodged in each of these are genes which cause femaleness to develop. There are many other genes linked with these female determinants that have nothing to do with sex. These genes are said to be *sex-linked*. The other 22 pairs of chromosomes are called autosomes, which probably carry, even in a woman, some male determining genes. Of course there are many thousands of other genes which are neighbours to these male determinants responsible for the development of characters other than sex. A baby boy and man have exactly the same number of autosomes—22 pairs—but only one *X* chromosome. To balance this lonely chromosome, is a small *Y* chromosome (see Plate 3*a*). At one time it was thought to have no significance at all in sex determination but it is now becoming clear that the *Y* chromosome is essential to maleness. The evidence is based on people who have one *X* chromosome besides the normal number of autosomes, the *Y* having been lost in the formation of an abnormal sperm. Such unfortunate '*XO*' people have, then, 45 chromosomes instead of the normal 46; they are feminine in appearance but remain immature and provide a confirmation of the male-determining powers of the *Y* chromosome. Another bit of evidence for this is provided by another human abnormality who carries in each cell 47 chromosomes instead of 46. This number is made up from 22 pairs of autosomes, two *X* chromosomes and a *Y* and the abnormality results from a sperm carrying *both X* and *Y* chromosomes or from an egg carrying *two X* chromosomes. These unlucky people look and behave like men but have small testes, are mostly sterile, and sometimes even mentally defective. Even though these men have, like normal women, a pair of *X* chromosomes, the action of the *Y* so overwhelms the female determinants on the *X* chromosomes, that a nearly normal male is produced. Such '*XXY*' people may occur once in about 500 births! A double dose of the *Y* chromosome (see Plates 3*a* and *b*) has been shown to be linked with violent or aggressive behaviour (see Chapter 24). It is almost as if these men have become too 'male'!

It is the father's sperm that has the main say in whether a boy or girl baby is born and it happens like this. Before reproduction, as already explained on page 4, the double sets of chromosomes divide to form half packs of 23 each. Since in the man the *X* and *Y* chromosomes form one pair each half set will contain either the *X* or the *Y*, made up to 23 by the 22 autosomes. Millions of sperms are formed continually, each carrying a half set of chromosomes. Half of these will be *Y* carriers and half *X*. The eggs of the woman contain half packs of chromosomes and all will be *X* carriers, since there were pairs of these in the full chromosome sets. Thus, there is only one kind of egg but two kinds of sperm which differ by an *X* or *Y* chromosome. This is, of course, apart from the abnormal gametes described above. If a *Y*-bearing sperm fuses with an egg in reproduction the double set of chromosomes is restored with one pair *XY* and a baby boy is initiated. On the other hand if an *X*-bearing sperm fuses with an egg the double set of chromosomes contains a pair of *X* chromosomes and a baby girl starts to form. In most people it is a 50/50 chance of having a boy or girl because the father's sperm carries *X* and *Y* chromosomes in a 50/50 ratio. But it is well known that some families consist of all boys or all girls. Why is this?

THE SEX RATIO

The answer is that we do not yet know but it is worth pointing out that in lower animals genes are known that alter the proportion of *X*- and *Y*-bearing sperms. For example, in the fruit fly a gene is known that causes the destruction of nearly all the Y-bearing sperms. It is possible that similar genes exist in man and could explain families of all boys or all girls. But taking the population as a whole, the 'either/or' mechanism of the distribution of *X* and *Y* chromosomes would seem to guarantee equality in the numbers of boys and girls conceived and born. This simple guarantee does not come up to expectations. For every 100 girls born 103 to 106 boys are born (i.e. the sex ratio is 100 : 103–106) and at *conception* the excess of boys over girls is even greater (probably 100 girls to 150 boys) and suggest that the *Y*-bearing sperms may get to the egg first. The excess of boys, however, is reduced by miscarriage during pregnancy and by stillbirths, so that the proportions given above are reached at birth. Up to the 1950's, in England and Wales, the proportion of girls to boys was slightly higher at adolescence owing to greater death rate among boys in the first year of life.

Twenty years ago there were two million more women than men. Now, although all babies have a better chance of survival, boy babies do proportionately better than girls and a new trend has started in which there are more young men than young women. So, the term 'surplus women', an ungallant phrase, no longer exists in the marrying years. But between middle age and old age the proportion of women to men increases so that at the age of 85 there are about two women to every man. Middle age is the testing time for the vulnerable male. For example, in England and Wales in 1958 the death rate among men between 55 and 64 was double that of women in this age group. There is no doubt that in this country during the past 100 years the increase in expectation of life for middle-aged women can be put down largely to the reduction in child-bearing. In the West men and women have gained a control over fertility through mass-produced contraceptives, and now many modern mothers, most of them married by the time they are 25, have had their families within ten years of marriage and have half their lives before them to bring up and enjoy their children and to do a job outside their home. The average Victorian mother spent a large part of her married life having babies and bringing them up. For example, in 1870–79, 61% of married women had five or more children compared with only 11% married in 1925. In India continual child-bearing takes its toll of women just as it did here a century ago and, as a result, in India there are more men than women. In England and Wales middle-aged men seem to be just as vulnerable to certain diseases as they were a century ago, and high death rates are kept up by occupational hazards—working in polluted air, cold weather, road accidents and accidents at work—and by being constitutionally weaker than women. This matter is dealt with in the next two sections.

SEX-LINKED GENES

Sex linkage has already been mentioned on page 26 and refers to those genes carried on the *X* chromsome which do not affect sex. There are about 60 known genes on the *X* chromosome of man. Some of these sex-linked genes are harmful recessives and, in the male, are not masked because there are no corresponding genes on the *X*'s partner, the short *Y* chromosome. Consequently, sex-linked diseases can show up in men and boys. In a woman, on the other hand, such a harmful recessive on one *X* chromosome, is normally masked by a dominant gene on the other *X* chromosome and she will only be a carrier of the disease and will not usually show it, or if she does, it will be a very mild form indeed.

When a disease is due to a recessive sex-linked gene (see Figs. 8 and 9) a man cannot pass on the disease to his son because he transmits his *Y* chromosome to him and not his *X*, which goes to his daughter. On the other hand, if a women 'carried' a disease-producing recessive on one of her *X* chromosomes, half her children would get this particular *X*. The

FAMILY TREES OF PEOPLE WITH DRINKING DIABETES OF RENAL ORIGIN

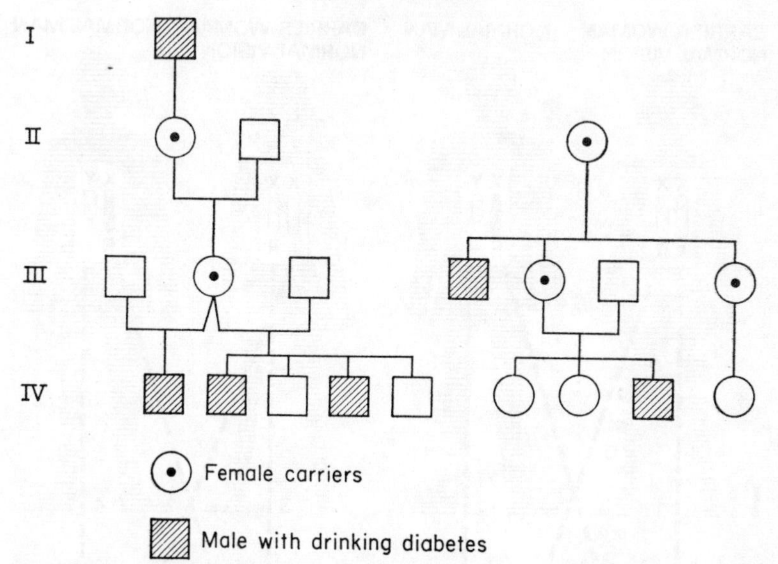

FIG. 8 Water diabetes is a disease in which large amounts of urine are passed due to a failure of the kidney to concentrate urine. It can be caused by a lack of hormone secreted by the pituitary gland and two genes, one dominant and one sex-linked recessive are thought to control it. A second type is more serious since unlike the first type it cannot be controlled. This second type is caused by a failure of the kidney tubules to concentrate urine even when the hormone balance is normal. A sex-linked recessive gene is thought to cause this 'renal' type. Carrier women can be detected because their ability to concentrate urine is usually well below that of normal people though better than that of the affected sons and brothers.

After Carter, C. O., *Human Heredity* (Pelican Books, 1962).

other half would get the normal *X*. If some of the children were boys, disease would show up in them but in girls the harmful recessive would be hidden and the girls would be 'carriers'. Queen Victoria was known to be a carrier of a serious blood disease, called haemophilia, in which the blood fails to clot after skin damage, or does so very slowly. As would be expected then, many male descendants of Victoria showed the disease; one son, at least three grandsons and six great grandsons have been haemophiliacs, including members of the late Royal Families of Spain and Tsarist Russia. Members of the present British Royal Family are free of the haemophilia gene. It is obvious from Chapter 1 that a single dominant or a pair of recessive genes can cause

SEX LINKAGE SHOWING THE INHERITANCE OF RED-GREEN COLOUR BLINDNESS

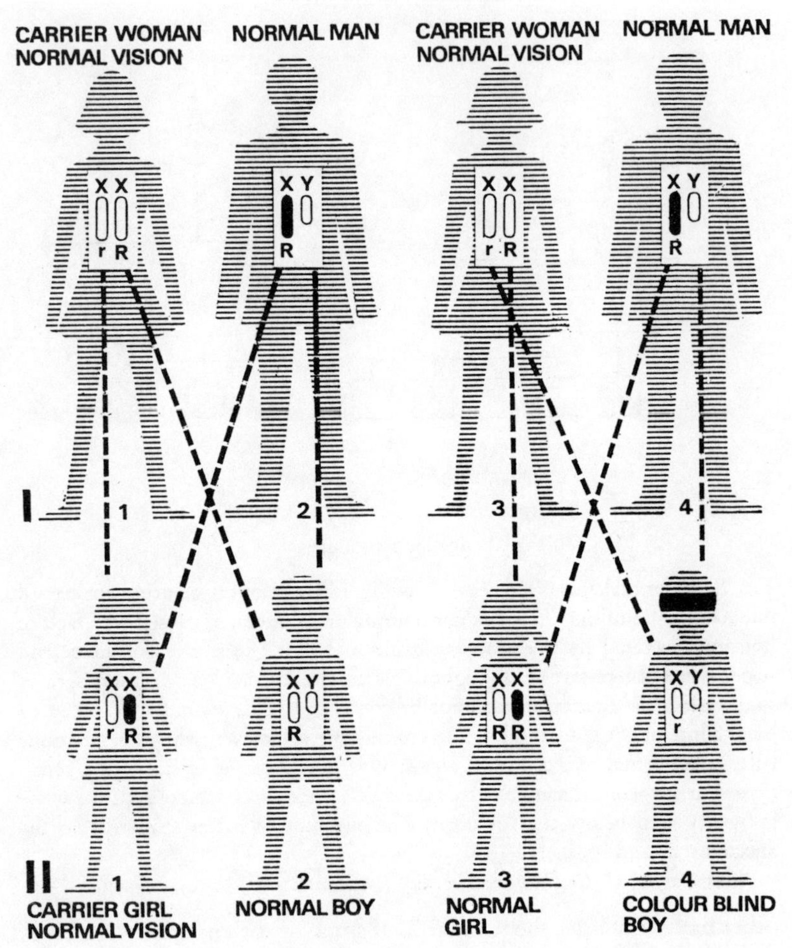

FIG. 9 There are around 60 known genes on the X chromosome (the best charted of all the human chromosomes) which have nothing to do with sex determination. Most of them are recessive and include the Duchenne type of muscular dystrophy; haemophilia; two different kinds of water diabetes, one due to kidney abnormality, the other to lack of posterior pituitary hormone; deficiency of eye pigment; porcupine skin—a rough scaly skin condition; some types of night blindness and red green colour blindness. The inheritance of the latter is shown in the diagram, but all the other conditions are inherited in the same way. In the above case a carrier woman I_1 and I_3 marries a normal man. If the X chromosome carrying the recessive gene goes to the girl (II_1) she will be a carrier like her mother because the recessive gene for the abnormal

much misery and suffering to individuals and families but in the case of haemophilia we have a gene that might have altered the course of history. In Russia and Spain the heirs apparent to both thrones were haemophiliacs and this fact might have had a favourable effect on the revolutions that led to the overthrow of the monarchies of both countries.

Another common sex-linked eye defect is red colour blindness where the affected person confuses red and green. It is caused by a recessive gene and affects about 2½% of men in England. The hypothetical family tree (see Fig. 9) shows that it is men who suffer from the defect and not women. Those who show the abnormality cannot do work which involves the recognition of signal lights; they are also unsuited to many branches of the paint, printing, textile and dyeing trades.

A type of muscular dystrophy, a serious muscle disorder, affects about 280 boys in every million born, and is due to a recessive sex-linked gene. Until recently, it has been impossible to tell whether a woman from an affected family is a carrier of the offending gene. If she is, half her boys would show the disease and half the daughters are in danger of being carriers like their mother. Now blood tests and examination of 'at risk' women can detect about 70% of carriers, some of whom have a mild muscle weakness and married couples can be told of the dangers of having children. It has been shown too that carriers of the haemophilia gene bleed for slightly longer than normal. In the future it may be possible to have certain tests before marriage to detect the presence of some recessive genes causing serious disease and then couples can decide, on the facts presented to them, whether to risk having a family.

---

trait will be partnered by a dominant gene on the father's X chromosome which suppresses the action of the recessive. If the recessive-carrying X goes to a boy (II_4) he will be colour blind because he has only one X chromosome and there is no possibility of a normal gene to suppress the action of the recessive. For a woman to be affected she must have the recessive gene on both X chromosomes which means that she must have received one abnormal gene from each parent, a rare event. Colour blindness is not a serious disease but many of the others mentioned above are. If one of these diseases runs in a family members should be told that if a carrier woman marries a normal man the probabilities are that the sons have an even chance of being normal or affected while her daughters have an even chance of being carriers of the trait or absolutely clear of it. Methods are being evolved now to detect carriers so that they can be warned.

THE WEAKER SEX

The male sex, then, has one *X* chromosome, most of whose genes have no partners on the *Y* chromosome. Consequently if these genes cause disease it will show up in men and boys only and we have seen that this is true for certain sex-linked defects and diseases. The number of stillbirths and infant deaths among boys is greater than among girls and this may be due to the action of 'bad' recessive genes harboured on the *X* chromosome.

Up to the age of thirteen, the muscular strengths of girls and boys are about the same, when measured by gripping, pulling and thrusting. After thirteen, the performances of boys and girls diverge and by sixteen boys are considerably stronger and bigger. These measurements were made on boys and girls who lived on similar foods and were given the same chances of developing their muscular strengths, by playing games. In short the male becomes at adolescence more adapted to such tasks as hunting, fighting and manipulating all sorts of heavy objects, as are necessary in some forms of food-gathering. Men are not only bigger but they burn up their food at a higher rate and consequently need larger quantities of food. It seems then that the differences in size and strength are more likely to be genetical than environmental. While there is obvious evidence that men are stronger physically than women, there is no doubt that women are tougher physiologically. In the affluent countries of Western Europe and America, the fragile men die more easily than women; they seem to have a lower resistance to infection than women, and bronchitis, lung cancer and heart attacks cut them down more readily than women in middle age. In this age group and in this country smoking, working in cold and smog, lack of exercise and emotional strain, add to the proneness of men to the diseases mentioned above (see also Chapter 13). Accidents on the road and at work add to the total. The general picture therefore is one of an increasing proportion of women to men at middle to old age and this may be due to both genetical and social influences operating in favour of women.

4: Mutation

MUTATION has been mentioned before in Chapter 2 and means a sudden and quite random but permanent chemical change in a gene which causes an effect of definite extent. Besides gene mutations there are other large kinds; those due to the addition or subtraction of complete or pieces of chromosomes. Mongolism in one or more of its forms (see Chapter 6) is due to the inheritance of a chromosome mutation. A famous example of a gene mutation, previously described in Chapter 3, is haemophilia. Such a mutation either took place in Queen Victoria or was passed on to her by one of her parents in whom it originated. A mutation that happens in a gamete does not show up in the person who formed that gamete. If it was a dominant mutation it would show its effect in the offspring (the *mutants*) of the individual in which the mutation took place. A recessive mutation would not behave in the same way. As described in Chapter 1 two recessive mutations have to be brought together by a sperm and an egg before the effect of the mutation is seen and rare recessive mutation may be hidden for generations before it meets up with a partner. However, a sex-linked recessive mutation (haemophilia) is usually spotted in one of the first generations after its occurrence, for in males a sex-linked gene shows itself at once no matter whether it is recessive or dominant.

SPONTANEOUS MUTATION

The causes of spontaneous mutation are numerous but there are three main causes.

1. The gene-copying process described in Chapter 2 may go wrong and produce a molecular structure different from that of the original gene; the original gene remains the same but it has produced a faulty copy of itself. The genes that cause albinism or haemophilia are faulty copies of the normal genes that enable pigment to form properly or blood to clot at the normal rate.

2. The natural 'background' radiation from cosmic rays or from soil and rocks (higher in granite than sandstone areas) or from chemical compounds in food may cause a change in a gene that is not dividing. Once it does start to produce copies of itself the error will be perpetuated until, by a further mutation, the normal gene is restored.

Most natural mutations seem to be due to a chemical or physical shock which shakes up and causes a slight alteration, or rearrangement, of the molecular pattern of the gene.

3. Chromosome breakages (see p. 47) are grosser types of mutation. There is evidence that these lesions, some just beyond the resolution of the microscope may be the cause of a great deal of congenital disease. Knowledge of how these breakages are caused is scant, but predisposing factors are exposure to radiation and increasing maternal age.

ARTIFICIAL MUTATIONS

In support of the chemical or physical shock theories we know that artificial mutations in flies, moulds and bacteria, can be produced by X-rays and chemicals like mustard gas. These mutations are no different from natural ones. Muller in 1927 showed that the rate of artificial mutation caused by X-rays could be stepped up in vinegar flies to a rate several hundred times as great as that found in nature but it is debatable whether the mutation rate in man can be increased artificially either by X-rays or by other types of radiation. Evidence from Hiroshima showed that while 0·89% of the children born to radiation-exposed parents were abnormal, 0·92% of those born to non-exposed parents were also abnormal. There is, however, some slight evidence to show that there was a rise of boy births where the fathers had been exposed to radiation and a decrease in boy births where the mothers had been exposed. While these results need cautious interpretation a possible explanation is as follows. In men exposed to radiation the relatively large *X* chromosomes (see Chapter 3) would be damaged more often than the smaller male-determining *Y* chromosomes. Moreover the *X* chromosomes would carry many more mutant killer genes than the *Y* which in any case has only a few genes to mutate. Thus more of the eggs which were fertilized by sperms containing damaged *X* chromosomes from irradiated fathers would die than of those fertilized by non-irradiated fathers, and the sex ratio would accordingly change in favour of boys. If women alone are exposed both *X* chromosomes may be damaged. Male zygotes (*XY*) would be destroyed by recessive mutant killer genes on the single *X* chromosome but female zygotes (*XX*) would be protected against recessive killer genes lodged in one *X* by favourable dominant genes on the other and the sex ratio would change in favour of girls.

Most of the radiation experienced by man, apart from natural radiation, comes from diagnostic X-rays whose benefit to humanity heavily outweighs their liability to cause mutation. But mutation there will be unless the reproductive organs are shielded; indeed some genetical damage has been found among the children of radiologists. The dangers of unnecessarily high doses of X-rays in medicine are more alarming than the testing of atomic bombs. And all-out atomic war would, of course, wipe out millions of people immediately and millions of others would be subject to chronic irradiation that would eventually kill. By these standards genetical damage appears a trivial matter.

If there was an increase in the continuous radiation to which the British population was exposed, through diagnostic X-rays, hydrogen bomb testing and the civilian uses of atomic energy, giving a 10% rise in the mutation rate, it has been calculated that the number of diabetics would be increased, after many generations, by about 25,000, congenital malformations by 10,500, schizophrenics by 6,900, deaf mutes by 1,500 and haemophiliacs by 100.

Perhaps more attention should be paid to the long-term effect of chemical compounds on mutation. Although many chemical compounds such as mustard gas cause mutation in bacteria, the same compounds may not be able to get through mammalian cell membranes to the genes in the reproductive organs. We do not know as yet which chemical substances cause gene mutation in man. It may be that drugs or chemicals present in food, or in insecticides sprayed on food, or nicotine, or caffeine are responsible. It is abundantly clear, however, that nicotine from smoking causes lung cells to produce tumours. If lung cancer is produced by chemicals in smoke acting upon the lung cells to produce *somatic* mutations (mutations occurring in body cells not concerned with reproduction), and subsequent tumour growth from them, then chemical *somatic* mutation is happening at an alarming rate.

There are still many questions to be answered about mutation. For example: Are the genes of men and women equally sensitive to radiation? What is the number of mutable genes in man? How does the sensitivity of human genes to radiation compare with that of the vinegar fly and the mouse?

FREQUENCY OF MUTATION: A RARE BUT IMPORTANT EVENT

J. B. S. Haldane calculated that haemophilia is kept in the English population by rare but spontaneous mutation. Perhaps a quarter of all haemophilia genes are wiped out by death in each generation. By a simple calculation from the existing frequency of haemophilia Haldane showed that if no new mutation was taking place, and haemophilia genes now existing were inherited from haemophiliac men and women of earlier populations, then at the time of the Norman Conquest all the men of England must have suffered from haemophilia. This, as Haldane pointed out, is nonsense and shows that fresh mutation at a rate of about one mutation in 50,000 individuals (i.e. one in 50,000 carries a new mutation for haemophilia) balances the loss of the gene by death of the carrier. This is a high rate of mutation. Other calculations show that for a certain tumour of the eye, the mutation rate of the normal to the abnormal gene is 0·4 per 100,000 gametes. Each individual gene, then, mutates very rarely but the number of genes in man must be very high. If, for example, there are 10,000 genes in a full set of human chromosomes, a mutation rate of one in 100,000 for each gene, in each gamete would mean that one gamete in ten carries a fresh mutation. Although this is a rough guess it is unlikely to be too high.

Even though mutation is rare it is very important, for it supplies the basic raw materials of evolution (see Chapter 9). Without mutation there would be a limited, though large, number of genetic constitutions but beyond this fixed limit no new and valuable combinations could arise. The effect of a fresh mutation is influenced by the presence of all the rest of the genes in a cell. So well balanced are the genes with the environment that only very rarely can a new mutation be an improvement; most of them are bad. A dominant mutation, which represents a big, violent change in the harmonious company of genes, would probably destroy the newly fertilized egg or perhaps cause the death of the carrier before he could perpetuate the mutation. Mutations with very small effects are more likely to be incorporated harmoniously among the genes and are likely to lead to an improvement of the stock. So important is the stability of the gene 'complex' that there are genes whose function is to control the rate of mutation of other genes. Too little mutation would lead to a species with a dead-end future lacking the power to adapt itself to a changing environment. Too much mutation, leading to wild variety, would produce a species unsuited to the existing environment. In short, a species will survive which fits its own environment and whose descendants will fit theirs.

5: The Origins of Human Individuality

WHAT are the causes of the likenesses and differences within a family? They may be due simply to the nature of the environment. Children learn from parents and sometimes follow the interests of the father in music or rowing or in his profession. Many sons of doctors and clergymen themselves follow these callings because they have grown up in a particular climate of work, talk, books and enthusiasm. The Bach family tree (see Fig. 10) is loaded with musicians (57* of them) and, certainly in the past, the chance of an eminent judge being the son of a judge was about one in three. All this may merely be due to living among certain people, but the potential to be a musician or a mathematician is likely to be inborn. It is very difficult indeed to disentangle the effects of heredity (nature) and environment (nurture) in a human individual. Basically, however, it can be said that a man's genetic endowment determines his potentialities and the environment determines the extent to which these potentialities will be fulfilled during development.

Just how difficult it is, in the case of humans, to unravel which charac-

* Not all shown.

PART OF THE BACH FAMILY TREE

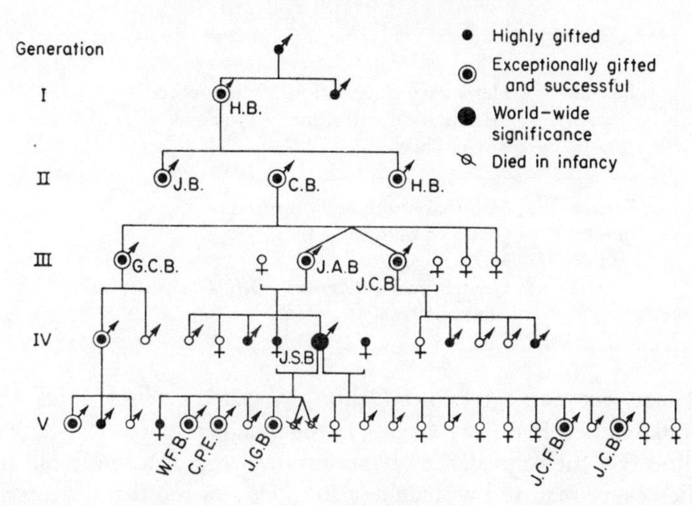

FIG. 10 Bach pedigree showing the inheritance of musical talent. Among the 20 children of J. S. B. not one was altogether unmusical. The twins Johann Ambrosius and Johann Cristoph were probably identical. They had similar musical talents, looked alike, had similar illnesses and died about the same time. The pedigree also may illustrate the effect of cousin marriage on fertility. J. S. B. married twice. His first wife, Maria Barbara, was a second cousin and bore him 7 children. The second marriage was to Anna Magdalena Wilcken who was unrelated to him and bore him 13 children. (See also p. 60.)

teristics are due to nature and which to nurture, is shown by a recent study by Professor Penrose of the causes of variation in the weight of babies at birth (see Table 2). It will be seen that 30% of the causes of weight variation remain unknown. Indeed, it would be remarkable if all the differences between human beings could be strictly determined.

It might be useful to recapitulate briefly here the main causes of inborn differences, those due to nature. Some of these have been discussed in detail in the previous chapters. First, there are various kinds of ancestral differences such as those between a man, a mouse and a bird on the one hand, and those between a Chinese and a Negro on the other. Secondly, there are differences due to the separation of pairs of genes in the formation of eggs and sperms and their chance recombination at fertilization to give rise to children bearing certain characteristics, some of them novel, in a definite ratio. Finally, there are differences due to mutation of genes or chromosomes.

TABLE 2

		%
Inherited factors	Hereditary constitution of the mother	20
	Hereditary constitution of the child	16
	Sex of child	2
Environmental factors	Maternal health and nutrition	24
	Order of birth	7
	Mother's age	1
	Unidentified influences, such as posture of foetus	30

Ninety-six years ago Francis Galton, a versatile genius himself, wrote a book entitled *Hereditary Genius*. In the final paragraphs it is clear that Galton saw the importance of the fertilized egg as a single cell from which every man and woman develops. He saw too that the interplay of unique environment and unique heredity, that is, nature and nurture, was the cause of the uniqueness of the individual. He says 'we may look upon each individual as something not wholly detached from its parent source—as a wave that has been lifted and shaped by normal conditions in an unknown, illimitable ocean'. How is it possible to make an estimate of the effects of differences in heredity and in environment that make men and women individual? It is relatively simple in plants because in them we can keep the heredity (the stock of genes they were endowed with) constant and alter the environment. If heredity is stronger than environment there will be little change in response to the changed conditions; but if they change out of all recognition we know that the changed conditions have over-ruled heredity. We might, for example, take cuttings from a plant and grow some in the light, some in the dark, in a warm, and in a cold place. The pieces of plant will all be the same genetically because they come from the same plant, only the environment will have been altered. But this obviously is not possible in man.

TWINS

Some experimental evidence about the causes of human individuality is provided by the study of twins. About one in every 88 maternities produces twins. Sometimes these develop from one fertilized egg

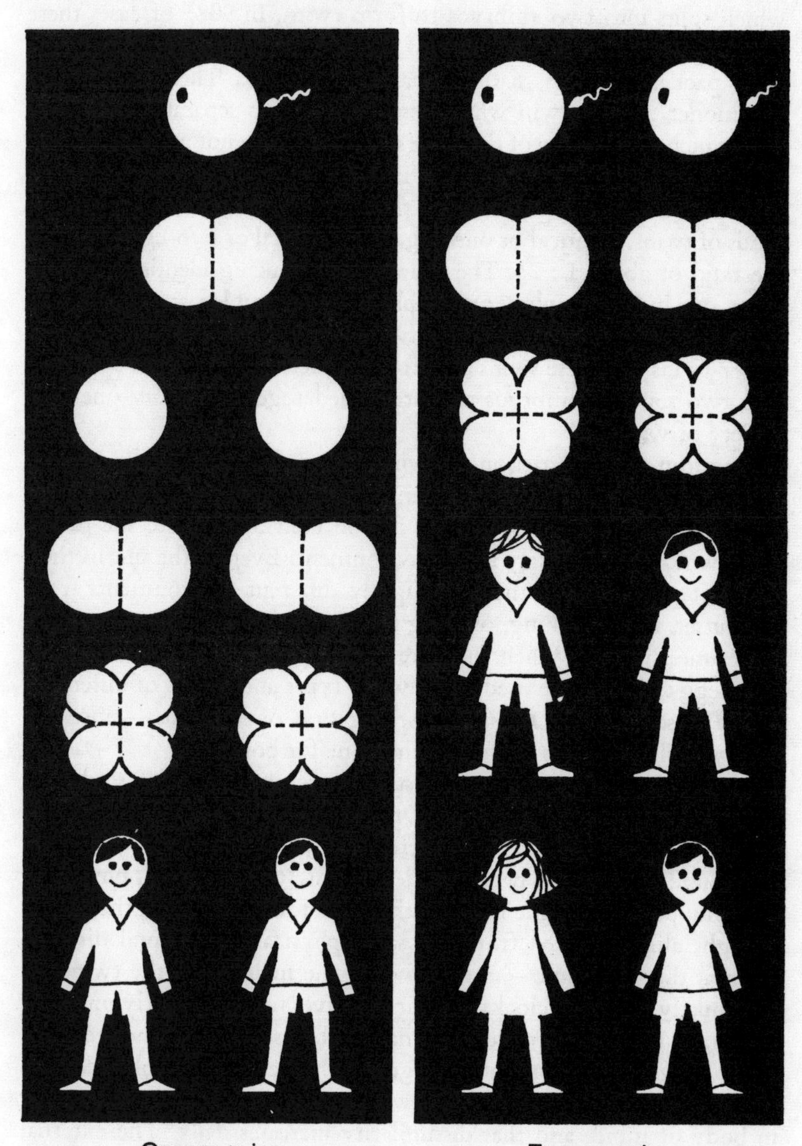

Fig. 11 Twins
After *Growing Up* (B.B.C. Science Work Units, Summer 1966).

which splits into two embryos to form twins. In 90% of cases these 'one egg' twins are identical; two duplicate human beings with the same packet of genes, nature, character and sex. There is another, commoner type of twin which results from two separately fertilized eggs which need not be of the same sex and need be not more alike than ordinary brothers and sisters (see Fig. 11). This is because they have only half their chromosomes on the average in common. The two kinds of twins, identical or one-egg, and fraternal or two-egg, occur in the ratio of about 1 : 2½. The chance of producing identical twins is about one in 300. Triplets, quadruplets and quintuplets are varieties of, or combinations of, one-egg and two-egg twins and have been called 'super-twins'. Siamese twins arise from *one* fertilized egg that splits late into two separate embryos and are joined together in any one of a variety of ways.

How can the comparison of twins be used to estimate the relative importance of nature and nurture? One-egg twins, since they have a common hereditary endowment, can be used to estimate the differences which arise entirely from environment. Even in the uterus their different positions put them in slightly different environments from the outset. More striking, one egg twins may be brought up apart, as sometimes happens when they are orphaned and separated at birth. Two-egg twins can be used to show the types and shades of difference which arise within a family by segregation of the genes—since the family environment is essentially the same for both.

Francis Galton made the earliest start in using twins to study the heredity-environment problem. One-egg twins he found were often mistaken for one another and were usually distinguished by ribbons tied round the wrist or neck, but despite this 'the one is fed, physicked and whipped by mistake for the other'. He found, too, that they often thought alike, suffered from the same physical and mental diseases, often at the same time 'in cases where the maladies of the twins are continually alike, the clocks of their two lives move regularly on at the same rate, governed by their internal mechanism'. Galton noticed that the two-egg twins brought up together showed marked physical and mental differences. In one case he wrote, 'they were never alike, either in body of mind, and that dissimilarity increases daily. The external influences have been identical.' To Galton these twin studies meant one thing, that physical and mental traits are chiefly determined by 'inborn nature' or, as we know now, the inherited factors or genes; environmental differences acting on identical heredities seemed to have little effect and different heredities overwhelmed the effect of similar upbringing and environment.

How do the results of Galton and his successors stand up to more modern research on twins? It has to be admitted that comparisons of one-egg and two-egg twins do not give the clear-cut separation of heredity and environment that Galton thought and hoped for. One-egg twins do not always have the same chromosome outfits. One may lose an *X* or a *Y* chromosome (see Chapter 3) as the fertilized egg splits, and change its sex. Moreover, if the cytoplasm round the egg splits unevenly one twin may lose a foot, or have a different temperament or intelligence from the other. Two-egg twins are formed from *whole* eggs so there are no cytoplasmic differences. Any differences depend on how widely their chromosomes differ. Sometimes they are very similar so that two-egg twins can be alike in many ways. Broadly speaking, however, although Galton's data were rather meagre and lacked objectivity, the conclusions he drew from them have been in most cases substantiated by later work, namely that differences in physiology, physique, intelligence and temperament are largely influenced by the internal character of the fertilized egg.

Some comparisons of twins have shown that some characteristics are strongly inherited and are quite unaffected by environment. These are the blood group and finger-print patterns. Other characteristics are almost as strongly inherited but can be slightly changed by the environment. Some of these are eye and hair colour, shape, length and colour of eyelashes and eyebrows, complexion, presence or absence or size and number of freckles, size and shape of nose, mouth and ears, teeth shape. Not only have recent twin studies thrown light on the inheritance of physical characteristics but they have shown that in susceptibility to disease and time of onset of disease, in mental and personality characteristics, one-egg twins show a greater similarity than two-egg twins, and this is true even when the identical twins have lived in totally different environments. Table 3 p. 44 summarises some of the evidence.

It would be foolish to suggest that twin studies alone can give a convincing picture of the origins of human individuality, especially in relation to intelligence and personality, but evidence can be brought to bear on these traits in a number of ways so that a convincing total picture begins to emerge about genetic psychology. The analysis of family trees of eminent men by Galton, described in Chapter 7, shows that the chance of an eminent man having an eminent relative is high. In the case of the Darwin–Wedgwood–Galton family (shown in

TABLE 3

Characteristics	*Concordance**	
	Identical twins (one-egg)	*Fraternal twins (two-egg)*
	%	%
Diabetes	84	37
Rickets	88	22
Tuberculosis	87	25
Cancer	58	24
Feeble-mindedness	94	54
Schizophrenia	68	11

* When a pair of twins both have the same peculiarity, it is said to be concordant, but when only one member of a pair has it, the pair are said to be discordant.

Fig. 12), for example, it is clear that high intelligence is transmitted and presumably such personality traits as energy, perseverence and imagination. Another piece of evidence about the inheritance of intelligence comes from correlational studies—the resemblance between parents and children and between the children themselves. Generally speaking, if both parents are of high intelligence a high proportion of the children are bright, and if one (or both parents) are of low intelligence the proportion of intelligent children is correspondingly less. A third piece of evidence is derived from adopted children (see Chapter 7) where the I.Q.s of children of professional men are six points above those of their adopted brothers even though they share the same home environment, while the I.Q.s of children of unskilled workers are about six points less than those of their adopted children. Does this mean that a social class is already printed on the child by heredity? A fourth piece of evidence is drawn from chromosome studies where one extra autosome (as we shall see in the next chapter) or a piece of a chromosome can cause mongolism which goes with mental backwardness while an extra *Y* chromosome is associated with violent anti-social behaviour. Even the inheritance of one recessive gene in double dose can cause mental backwardness in babies unless their diet is altered. Fifthly there is the mounting evidence on the importance of heredity in the inheritance of personality and temperament traits in rats described in Chapter 30. And

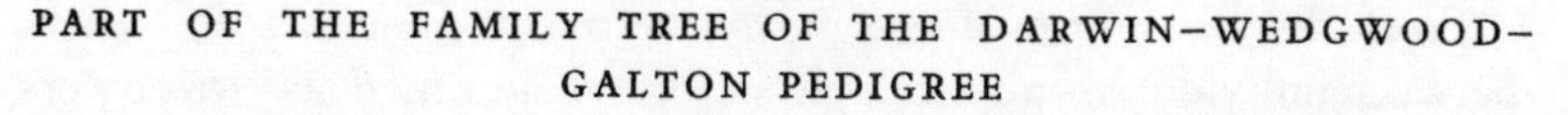

PART OF THE FAMILY TREE OF THE DARWIN–WEDGWOOD–GALTON PEDIGREE

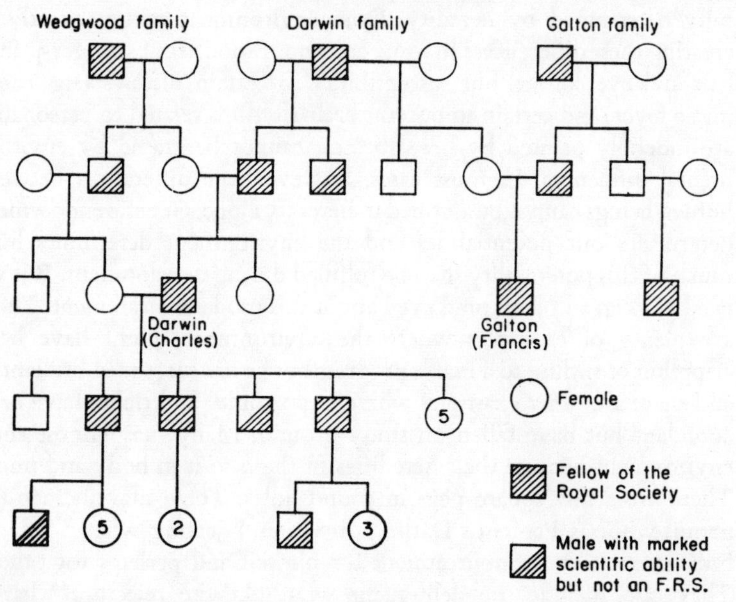

FIG. 12 This family tree shows the inheritance of scientific talent. Presumably women were also highly gifted, but in existing social circumstances had less opportunity for manifesting their endowments. Galton pointed out that the early environment may have been important in bringing out the talent in great scientists. He believed that an unusually large proportion of the sons of the most gifted men of science became distinguished because they had been brought up 'in an atmosphere of free enquiry'.

we shall see that race crossing between gypsies and 'whites' results in genetic recombination followed by a genetic determination of be-behaviour. Finally there are more modern studies on twins and the evidence they provide for the inheritance of criminal tendencies, mental illness, homosexuality and alcoholism. But a vast field still remains to be mapped before human individuality is well understood and twin studies offer opportunities for this. Particularly fruitful might be studies on the electrical activity of the brains (see Chapter 28) of one-egg and two-egg twins and its relation to crime and human behaviour in general. Already it has been shown that certain brain rhythms are as constant as finger-prints in identical twins and are inherited just as strongly.

From this body of evidence we are bound to infer that differences between individuals—physical, physiological, and mental—are powerfully determined by heredity. Does environment matter greatly in creating these differences? In some cases no. Blood groups, finger-prints, hair and eye colour, our susceptibility to certain diseases (e.g. rheumatic fever) and certain important brain rhythms related to personality are indelibly printed by heredity and cannot be erased by environmental influences. In most cases, however, the differences between human beings cannot be ascribed to heredity alone. Genetic endowment determines our potentialities and the environment determines how much of this potentiality shall be fulfilled during development. But we need to keep an open mind even about this moderate statement. There are plenty of examples where the environment might have been crippling or stifling to a man's progress but he has succeeded brilliantly, and there are other examples where opportunity and riches have been abundant but have fallen on stony ground. Many may choose their environment because their heredities fit them to it in body and mind. There are a few square pegs in round holes. These may include the genius who, as Professor Darlington writes, rejecting what is offered him, creates a new environment for himself and perhaps for others. They also include the delinquent who, likewise rejecting what is offered him, finds it easier to destroy than create.

6: Chromosome Mistakes

EXTRA chromosomes and their effects on sex have already been mentioned. It is remarkable that the automatic mechanism of cell division so rarely deals a bad hand in gamete formation. Almost always, exact half packs of chromosomes are dealt into each egg or sperm. When this happens, normal eggs and sperms are produced, each with the correct number of autosomes and one sex chromosome; where it does not, as we have seen, an egg might get two *X* chromosomes instead of one or a sperm might get an *X and Y* chromosome to produce the abnormal men and women mentioned in Chapter 3. The frequency *at birth* of babies with chromosome errors *XXY*, *XXX*, and *XO* (*O* meaning without another *X* or *Y*) is about one in 750, one in 1,500 and one in 5,000 respectively. But all the evidence points to a much higher frequency of *sex* chromosome mistakes which lead to early abortion and stillbirth. That is they never see the light of day.

Sometimes an extra *autosome* gets into an egg so that when it is fertilized by a normal sperm, 47 instead of 46 chromosomes are found in the cells of the offspring (see Plate 4). This extra chromosome, a third 21★ may be responsible for producing a mongol child. Mongol

★ See Plate 3*b* for classification of human chromosomes.

children are physically dwarfed and mentally backward and often have heart defects. Only about half survive their first year of life as they are extremely susceptible to infection. The frequency of mongol births in Western Europe and America is about one in 700 live births; the chance of having a mongol baby is greatly influenced by the mother's age, while the father's age is unimportant. The chance of a mother of twenty having a mongol baby is one in 3,000; of 25, one in 2,000; of 40, one in 100, and of 45, one in 50. It looks as if the mechanism for the *exact* division of the chromosomes wears out with age.

We saw in the last chapter the fine-grained but drastic effects of recessive genes on the brain. Here we have the much coarser effect of an extra chromosome that seems to upset the normal balance of genes and chromosomes to produce drastic physical and mental abnormality. Indeed about 20% of spontaneous abortions, 11% of sub-fertility in men and about 6% of mental defectives in institutions have chromosome mistakes. All in all about one in 200 newborn babies has an error of development due to a chromosome abnormality big enough to be seen under the microscope. It may be possible in the future, to spot not only extra chromosomes but bits of chromosomes that have 'chipped' off one and become stuck to another (translocation). This effect is known to cause about 5% of all cases of mongolism and may well cause other diseases and physical abnormalities too. In the case of translocation mongolism, in one family, a piece of chromosome 21 attached itself to chromosome 15. Thus the mongol child had a pair of chromosome 21 and a pair of chromosome 15 one of which was made longer than the other by a piece of a third 21.

7: The Inheritance of Genius

EXCEPTIONALLY, a genius is born to unexceptional parents. This is because he has inherited an arrangement of genes which, on the basis of the chance shuffling of chromosomes and genes at gamete formation, is extremely improbable. Shakespeare, Goethe, Beethoven, Newton, Ampère, Charlie Chaplin, Dickens, Gauss, Faraday, Kepler, and Einstein appeared out of the blue, although it is likely that their parents were intelligent. The chance of another Shakespeare turning up in the same family, for instance in a son, is highly unlikely because when the genius reproduces his particular gene arrangements are split up. Although they are the same genes that are transmitted to the son, it is even more unlikely that the gene combinations that helped him to become a genius will emerge exactly the same in the child. On the whole children are just about as much like their parents in intelligence as they are in height. However, it is more likely that an extraordinary child will be born to gifted parents than to ordinary parents. Francis Galton in his great book, *Hereditary Genius*, analysed the inheritance of high ability in judges, statesmen, commanders, literary men, scientists, poets, musicians and painters, and came to the conclusion that mental ability is strongly inherited but

PERCENTAGE OF EMINENT MEN IN EACH DEGREE OF KINSHIP TO THE MOST GIFTED MEMBER OF DISTINGUISHED FAMILIES (JUDGES)

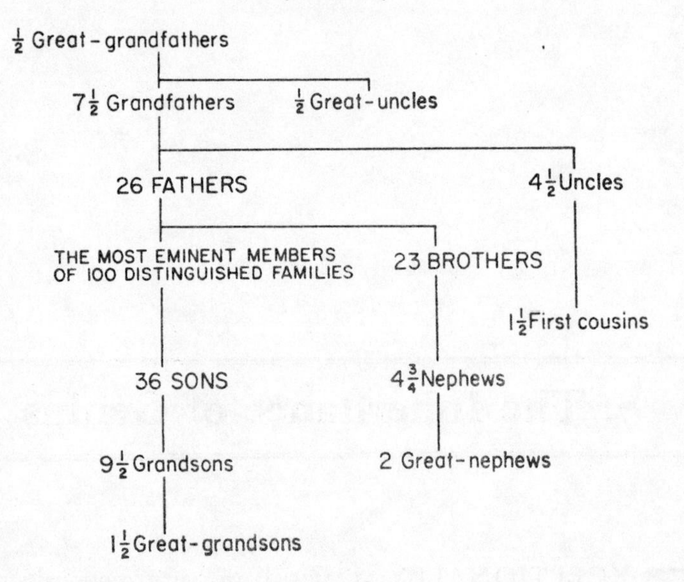

FIG. 13 The table shows in the most unmistakable manner the enormous odds that a near kinsman has over one that is remote, in the chance of inheriting ability. Speaking roughly the percentages are quartered at each successive remove, whether by descent or collaterally. Thus in the first degree of kinship the percentage is about 28; in the second, about 7; and in the third, 1½.
After Galton, F., *Hereditary Genius*, (Fontana edition, Collins, 1962).

found that the chances of an outstanding man having outstanding relations decreases as the relations become more distant. The effects of heredity are, as it were, watered down. For example, only 26% of the fathers of eminent judges were eminent, 7·5% of grandfathers, and 0·5% of great grandfathers but 36% of their sons are eminent (see Fig. 13).

Occasionally long family trees of exceptionally gifted people show the perpetuation of gifts as well as the step by step assembly of an outstanding gene combination culminating in a genius such as Charles Darwin, Francis Galton, John Sebastian Bach or Titian (see Figs. 10 and 12). But the genius of Erasmus Darwin turned up out of the blue.

The maintenance of high intelligence in a family like the Darwins depends to a large extent on the eminent marrying the eminent and in general this happens. The dull marry the dull and the gifted the gifted because they take pleasure in each other's company. Galton noticed this and pointed out that his particular eminent man met his future wife in the society of his own friends and she was therefore not likely to be stupid! She was usually related to some of them and therefore had 'a probability of being hereditarily gifted'. Philip II of Macedonia, illustrious general and statesman, married Olympias and fathered Alexander the Great; Goethe's mother was a vivid and pleasant figure in German literature; the great Francis Bacon married an accomplished daughter of Sir Anthony Cooke, and so on. This custom of like marrying like, called *assortative mating*, is one of the foundation stones on the structure of our society and is discussed in Chapter 8. It fosters marriages between people with similar interests and characteristics such as a common religion (the Jews and the Quakers, for example), a similar standard of education and equality of social standing. And because of assortative mating society is split into myriads of groups.

THE ENVIRONMENT AND GENIUS

Looking at the family trees in Figs. 10 and 12, it looks as if ability is linked with the male sex for very few women appear in the distinguished lines. Presumably (and as Galton has shown in some cases), women were also highly gifted but in existing social circumstances had less opportunity for showing their natural endowments. While there is evidence that special abilities in music and mathematics and general intelligence are inherited, living in a home where there are books, good conversation and an opportunity to meet stimulating people helps the genes to act. The family tree of the Bachs is loaded with famous musicians and although the genes were there all the instruments were on tap too and so was expert teaching from an early age as well as the prospect of good employment. Without the genes no amount of practice would make a Bach but with no instruments or interest and encouragement from parents even the potentialities of a John Sebastian might have been snuffed. Galton knew of this in the case of scientists. He showed that the number of eminent fathers of the great scientific men compared with that of their sons were 26 and 60 respectively whereas the average of all the groups of eminent fathers and sons gave

31 and 48 respectively. This told him that scientific men inherited much from their mothers both genetically and through training. It also told him that the first in the family, like Erasmus Darwin, who had scientific gifts, was not nearly so likely to achieve eminence as the descendant who was taught to follow science as a profession and 'not waste his powers on profitless speculations'.

STUDIES ON TWINS AND ADOPTED CHILDREN

To pass on to less rarified but no less important studies of 'intelligence'. This is the rather narrow quality measured by the highly fallible method of intelligence tests, the conventional ones, not those which test 'creativity'. The measured intelligence is called 'I.Q.' for short. The important thing to remember here is that a person's I.Q. is no use by itself. It only has meaning in relation to other people's I.Q.s in a population.

I.Q. scores on one-egg twins brought up together and others that have been reared in different environments show that the I.Q.s of the former are much closer, varying between 0 and 9 points, while the range of variation between the latter can be as great as 20 points. The normal variation in a population spans at least 100 points (an average person is supposed to have an I.Q. of 100) so, despite environmental effects, the genes seem to tether the individual within fairly close limits. Supporting evidence for this comes from a comparison of the I.Q.s of the adopted and natural children of people with different occupations (see Table 4).

In this study all the adopted children were placed in foster homes before the age of six months and had an I.Q. test between the age of five and fourteen. It will be seen that there is a difference of about 17 points over the range of natural children where genetic as well as environmental influences are interacting but only about five among the adopted children with only environmental differences at work. The I.Q.s of the children of professional men are six points *higher* than those of their adopted brothers even though they share the same home environment, while those of the natural children of unskilled workers are about six points *less* than those of their adopted children. All this evidence points to the importance of genetic endowment. Another study of I.Q.s of children from the professional classes in orphanages shows that their I.Q.s are slightly higher on average than the rest, again

TABLE 4

A comparison between father's occupation and intelligence-test score of natural and adopted children. (From C. O. Carter, Human Heredity, *Pelican Books.)*

Father's occupation	*Adopted children*		*Own children*	
	No.	*Average score*	*No.*	*Average score*
Professional	43	112·6	40	118·6
Managerial	38	111·6	42	117·6
Clerical & skilled manual	44	110·6	43	106·9
Semi-skilled	45	109·4	46	101·1
Unskilled	24	107·8	23	102·1

possibly reflecting a genetical difference. But we need to be cautious. The genetic imprint seems to override the effects of the environment but we must remember that I.Q. tests are planned by intelligent professionals and thus there may be bias.

DECLINING INTELLIGENCE

In recent years voices have been raised which say that because the manual workers, whose children have a lower average I.Q. than the children of the professional and managerial classes, had larger families on the average than the rich, then it follows that the proportions of those with high I.Q. will steadily go down. On the surface this sounds a good argument. It is true that there is a relationship between intelligence and family size. The larger families which contribute most members to the population are in general of slightly lower intelligence and would appear to be swamping the smaller higher I.Q. families of the professional and managerial classes. The facts from I.Q. tests seem to contradict this. In 1932 and 1947 the I.Q.s of a large number of Scottish eleven-year-olds were measured and instead of a decline, as predicted, there was a slight increase! Out of a maximum score of 76 points the average in 1932 was 34·5: in 1947 it was 36·7. The results have had no convincing explanation and are a paradox. An interval of 15 years seems too short to show up genetic changes, perhaps the apparent rise in intelligence was due to an increased skill in doing intelligence tests and masked a real decline. Perhaps, too, better environmental conditions, food, health and education might have caused the rise. But there are too many factors varying to make one test significant.

8: Marriage and Close Marriage

WHOM DO PEOPLE MARRY?

PREVIOUS chapters have been concerned with the quintessence of heredity, the distribution of chromosomes and genes down the stream of generations. An ever changing pattern like the interplay of light on water, as the genes mingle and intermingle, toss up striking novelty here and there, submerge and reappear, but in general maintaining a degree of stability from generation to generation. But, for different gene combinations to meet and mingle, people must marry and the way in which they meet is of great biological and social importance. Selection of a spouse is not at all haphazard and the choice is made *within certain limits.* The first and most obvious limitation is that people choose their spouses from among people they meet, at work or at university, or from among people they have grown up with in the same street, or village or town. The second is not concerned with geographical distance but with social distance; there is still a frozen waste between 'classes' kept at a low temperature by the keen winds of religious and educational differences. Third, most subtle and least understood, is the attraction between people of similar physical and psychological characteristics.

THE BASIS OF SOCIETY

These attractions and barriers act to funnel people of common origin, education and religious background and foster like to marry like. This process of *assortative mating* has already been mentioned and is one of the most powerful separating forces in society and leads to the formation of groups, cliques and classes. Between these groups and classes marriage is discouraged but within them it is encouraged. Galton's eminent men married into an eminent class, Jew marries Jew, and Quaker, Quaker. The caste system of India divides the Hindu community into some 2,300 intra-marrying groups. These vary in size from a few hundreds to a hundred thousand or more and some have been going for 50 generations and some for one. In this country in the past, difficulties of transport greatly limited the choice of marriage partner for the majority but now ease of travel can bring 'foreigners' to town or village and with them new gene combinations. For example, in some rural parishes of Hertfordshire there has been a 50% fall over the last century in marriages between people from the same or the next parish—the motor-bike, as J. B. S. Haldane said, has been responsible for a great deal of genetic mixing! Communities vary in their stability.

In Bethnal Green ten years ago, 64% of married couples had been born there and had similar backgrounds and in Aberdeen 80% of spouses were born and bred there while 2% of marriages were between first cousins. In over a third of these marriages, the couples had known each other since childhood, perhaps living near to each other or going to the same school and had the same social, religious and educative background. But 20% of Aberdonians marry outsiders and this ensures a considerable mixing of genes. Being born in a place, marrying the neighbour's son or daughter, pushing a pram and dying in the same place still holds true for certain groups, for example among certain unskilled workers whose families tend to live for generations in the same district, perhaps leading to a clustering of genetic traits among them. In Britain the greatest *geographical* movement is among the small group of professional people with scientific, technical and administrative qualifications. But the greatest *social* mobility lies in the groups of skilled workers who percolate up and down the classes; upwards, generally, in the cities where skill and specialization are required. Women have always been able to bound across classes because of the adaptability of their manners and accents and because of men's interest in their looks, and certainly more women than men tend to marry into a class above their own.

Faster, easier travel and especially the spread of education at all levels, are opening up the 'classes' and widening the marriage choice for everyone, leading to shuffling of the genes. Education is far more important than social background in the choice of a spouse and this is obvious among people with a university or university type education. In one survey it was found that 71% of spouses had similar educational backgrounds but only 45% had similar social backgrounds.

Professor C. D. Darlington has given a great deal of thought to the biological importance of stratified societies like our own. He suggests that the societies that have survived have allowed a limited amount of intermarriage between the stratified groups. Intermarriage within each group preserves their genetic structure but between different groups old and tried gene combinations are taken apart and reassembled to produce new genetic variation in children. Too much of this new variation would not fit progeny to existing environments but might give them the genetic potential to overcome drastic cultural change. Nations that have survived for long periods, like the Romans under the Empire, have by the rules and conventions governing intermarriage permitted only limited intermarriage between classes. Such a percolation of genes up and down the different strata of societies fits the bulk of the population to the existing environment by maintaining a reservoir of old and tried gene combinations but at the same time allowing some new ones to be tried out. A few of these give types that are usually defective and eccentric and often infertile but some might have the versatility, originality and fertility to sustain drastic cultural change. For survival then, populations require adaptation, genetical stability and versatility. And the rules and regulations governing marriage and intermarriage (and immigration) in a society control its genetic structure and to some extent its future, although man can alter his environment to some extent to suit his potentialities (see Chapter 9).

COUSIN MARRIAGE

The danger of cousin marriage is discussed in nearly every family in Europe. It is an old fear springing from an old rule of conduct. In the eleventh century the Catholic Church tried to stop cousin marriage between any kin nearer than what was called the seventh degree of kinship. Moreover the seventh degree was taken to mean that there must be no common ancestor within seven generations. Because of this rule William the Conqueror did wrong when he married Matilda, his second cousin once removed, a sin that was expiated in building the Abbaye aux Hommes and the Abbaye aux Dames at Caen. These rules, of course, have been broken repeatedly by dispensation and first cousin marriages have been widespread in Catholic communities up to the present day and are most frequent among royal marriages where uncle–niece marriages also took place until 1850. The Protestant churches gave up the nominal ban on cousin marriage after the Reformation but then began to enforce a ban on uncle–niece marriages.

History does not tell us what was feared in cousin or other close marriage; it might have been sterility, but most people now believe that such a marriage does give an increased risk of producing an idiot child or one defective in some other way. This belief is true in large communities where cousin marriage is rare, such as among Protestants in England (about 1% of marriages are between first cousins in England). But in some small communities such as religious minorities where cousin marriage is frequent, people take the opposite view and say that no increased frequency of defects is to be found in the offspring of cousin marriage. Professor Darlington writes

> certain kinds of rare defective conditions are much commoner in the offspring of cousin marriages in communities where such marriages are the exception. In small inbred communities where such marriages occur regularly, the situation will be different, for the defects will appear more quickly after their origin by mutation and will thus be the more quickly eliminated by the lower viability or lower fertility of those who show them. The community will, therefore, become, as we may say, adapted to inbreeding.

A ONE IN EIGHT CHANCE

In a marriage between people related by blood, it is obvious that the chances of both spouses carrying the same recessive genes is increased. In fact first cousins, on the average, have about one eighth of their genes in common. Supposing that one cousin, although completely normal, carried a recessive gene for phenylketonuria or albinism; the other cousin would have a one in eight chance of carrying the same gene and a marriage between them would have a one in four chance of producing a defective child. One pair in twenty of the parents of children with phenylketonuria are first cousins and one in five albinos are the result of first cousin marriages. The chances of random marriage bringing together a pair of recessive genes (and therefore producing a defective child) for phenylketonuria or albinism are about one in 25,000 and one in 20,000 respectively. Or, to put it another way, if a husband carries a particular recessive gene, say for albinism or phenylketonuria, the chances of his unrelated wife carrying it would be 1 in 70 or 1 in 80 respectively. But for first cousins, once one carries the gene, no matter how rare it is, the chances that the other also carries it are always the same—one in eight—a very much higher risk.

COUSIN MARRIAGES AND FERTILITY

Professor Darlington has given us clear evidence about the relationship between cousin marriage and the number of children such marriages produce. He investigated the fertility of persistently *inbred* communities like 'Mennonites', Quakers, Jews and Royal Families where it is relatively common to marry a first or second cousin, and also particular families like the famous Wedgwood family where cousin marriage was common. He also collected information from our own normally *outbred* society where cousin marriage is rare. The data he obtained are summarized in Table 5 in which the cousin marriages in *outbred* families are contrasted with the cousin marriages in *inbred* families. Children, grandchildren and great-grandchildren are given.

TABLE 5

Cousin marriages	*No. of families investigated*	*Children*	*G. Children*	*G.G. Children*	*Increase*
Outbred	39	234	342	499	×2·1
Inbred	9	57	146	257	×4·5

It will be seen that the descendants have multiplied in the two generations by 2·1 and 4·5. This result seems to show clearly that the effect of cousin marriages was to lower the fertility of the children in families where there had been previous unrelated marriage. Explanations of the low survival and fertility rates in families included insanity, blindness, tuberculosis, diabetes and epilepsy in men, and miscarriage in women. Not only do these clearly defined causes contribute to low fertility but also the vaguer, but no less important elements of fertility, such as the absence of the love of children or sexual desire, which are presumably genetically determined. But in normally inbred families where it is common for cousin to marry cousin no such evidence for low fertility is found. It is the change in the normal marriage pattern that produces the lowering of fertility, outbred switching to inbred and *vice versa*. To test this idea Darlington looked for examples of families in which a man or a woman married in succession two spouses, one a cousin and the other unrelated. The number of descendants and their behaviour could then be compared in each succeeding generation. The Bach family is known to have had no previous cousin marriage and was therefore a family 'used' to outbreeding; the dying-out of the second cousin line and vigorous continuance of the outbred (unrelated) line is historically the best known of its kind (see Table 6).

TABLE 6

John Sebastian Bach (1685–1750)
Maria Barbara Bach (1684–1720) 2nd cousin
Anna Magdalena Wilcken (1701–1760) unrelated

Marriage	*Date*	c				gc				ggc	
		t	s	m	wi	t	s	m	wi	t	
2nd cousin	1707	7	3	2	2	6	3	0	0	0	
Unrelated	1721	13	6	3	2	6	4	4	3	10	*

* 28 gggc; 52 ggggc. *Abbreviations:* t — total born alive
s — survived
m — married
wi — with issue
c — children
gc — grandchildren
ggc — greatgrandchildren

Professor Darlington has shown by the analysis of families which switch their normal pattern of marriage from related marriages to unrelated or from unrelated to related, like the Bach example above, that the break from the normal pattern also breaks the fertility. Some-

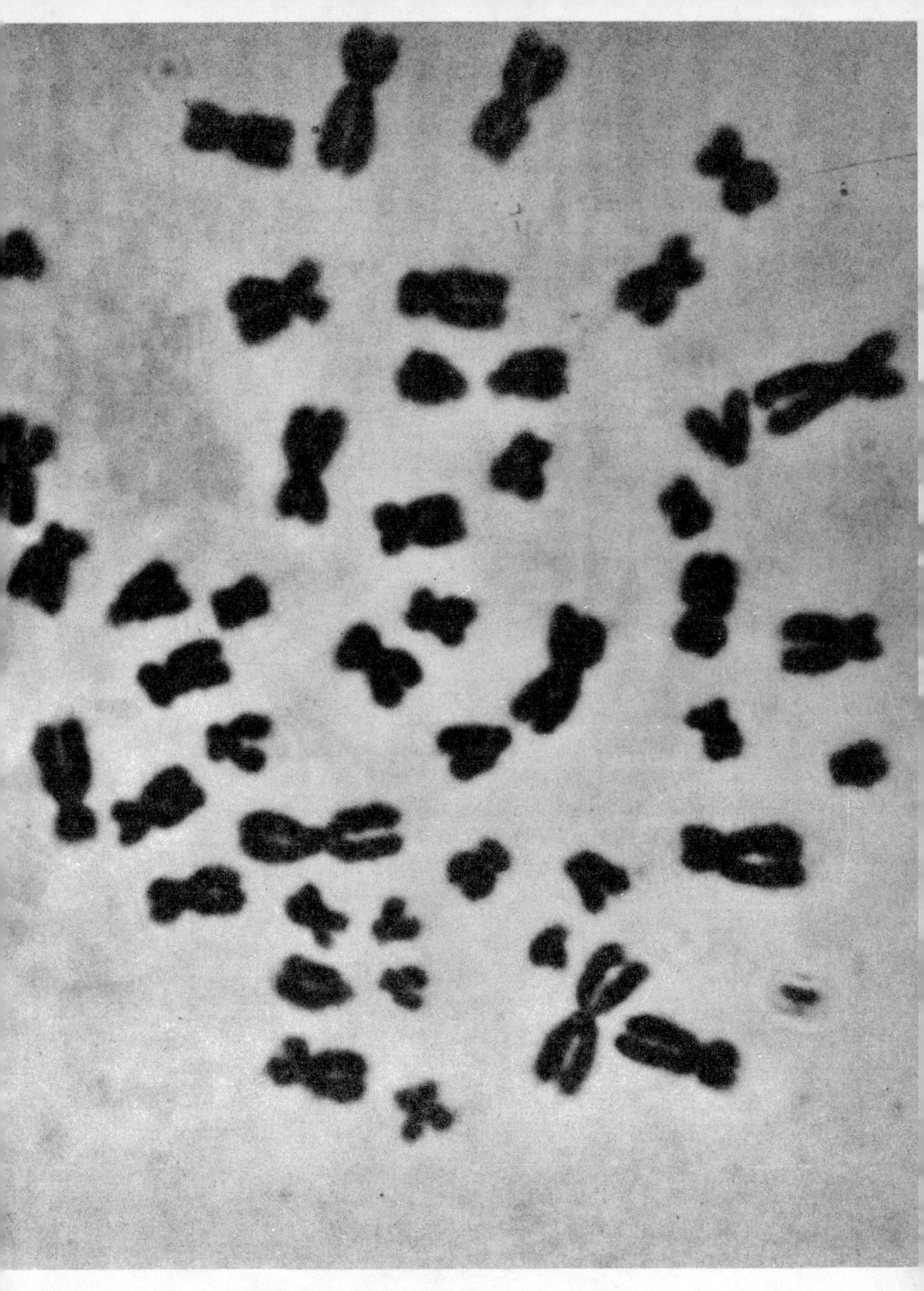

PLATE I—The mitotic complement of a normal human female showing the 46 chromosomes. × 5,500.

Preparation by courtesy of Dr Peter Pearson. Photo by Dr A. Haque and Mr C. G. Vosa, Botany School, Oxford.

Emperor Maximilian I
1459–1519

Margaret of Austria
1584–1611

PLATE 2—Hapsburg family

The narrow, undershot lower jaw and jutting lower lip can be traced back to the fourteenth century with the Hapsburg family. The trait is likely to be controlled by a single dominant gene.

Courtesy: Radio Times Hulton Picture Library.

Philip IV of Spain
1605–1665

Alfonso XIII of Spain
1886–1941

times the break leads to near extinction or actual extinction in later generations.

THE EXTINCTION OF GREAT FAMILIES

Why do the families of great men tend, with unusual frequency, to die out? In *Hereditary Genius* published in 1869, Sir Francis Galton investigated this problem by his analysis of the fertility of the English peerage. Of 31 peerages granted to the Judges of England who last sat on the bench near to the close of the reign of George IV, 12 became extinct and 19 remained. To explain these striking facts Galton lit upon a hypothesis which was 'simple, adequate and novel'. It was that peers and their sons like to marry heiresses and *vice versa*—perhaps because the peers needed money and the heiress wanted a title, although there may be other less obvious reasons. An heiress is usually the only child of a marriage and it is to be expected that she will not be so fertile as a woman with many brothers and sisters. This inherited lack of fertility can lead to the extinction of a title. In fact out of the 12 peerages that failed in the direct male line no less than eight failures were accounted for by heiress marriages. For example, the first Lord Colpepper (1664) married twice and had children by both marriages: in all, five sons and four daughters. The eldest son married an heiress and died childless. The second son married a co-heiress and had only one daughter. The third married but had no children and the other two never married so the title became extinct. The first Earl Cowper (1718) married twice. His first wife was an heiress and she had no children. His second wife had two sons and two daughters. His eldest son married a co-heiress for his first wife and had only one son and one daughter. The direct male line continued.

After ransacking Burke's volumes on the extant and on the extinct peerages Galton added the heiress marriages of judges to the heiress marriages of statesmen and Premiers, all of whom had been made peers, since George III came to the throne. He then was able to make comparisons, which are summarized in Table 7 (p. 62).

Galton found that among the wives of peers—

100 who were heiresses had 208 sons and 206 daughters;

100 who were not heiresses had 336 sons and 284 daughters.

He pointed out how exceedingly precarious must be the line of descent from an heiress. One fifth of heiresses studied by Galton had no

TABLE 7

Number of sons to each marriage	*Number of cases in which the mother was an heiress*	*Number of cases in which the mother was not an heiress*
0	22	2
1	16	10
2	22	14
3	22	34
4	10	20
5	6	8
6	2	8
7	0	4

N.B. 50 cases of each were taken which are given above but the results were doubled to turn them into percentages.

boys at all; a full third had only one child and three-fifths had only two. The salvation of many families came when the husband outlived the heiress, re-married and had children by his second wife.

On this evidence the heiresses with their inherited low fertility seemed to be the prime culprits in the extinction of great families. Even greater support was lent to this suggestion when Galton examined the extinction of dukedoms, earldoms, and baronies.

Every advancement in dignity is a fresh inducement to the introduction of another heiress into the family. Consequently, Dukes have a greater impregnation of heiress blood than Earls, and Dukedoms might be expected to be more frequently extinguished than Earldoms and Earldoms to be more frequently extinguished than Baronies. Experience shows this to be most decidedly the case. Sir Bernard Burke in his preface to the *Extinct Peerages** states that all the English dukedoms created from the commencement of the reign of Charles II are gone, excepting three that are merged in royalty, and that only eleven earldoms remain out of the many created by the Normans, Plantagenets and Tudors.

Would the inherited low fertility of heiresses be enough in itself to snuff out these great families? Let us go back to the idea referred to earlier in this chapter, of assortative mating—like marrying like, and the attraction of people of similar physical and psychological characteristics. It is possible, according to this theory, that Galton's heiresses of low fertility married peers of low fertility. The peers who married

* See p. 186, *Hereditary Genius*, Fontana, 1962.

heiresses may not have been so interested in children or having children as the peers who did not marry heiresses and this type of assortative mating, adding together infertility, would tend to wipe out families. The theory is equally applicable to peers or ploughmen if the idea of like marrying like holds good; it would be true for people of high fertility or medium fertility marrying each other as well as those of low fertility.

9: Natural Selection in Man

EVOLUTION by natural selection is the essence of the Darwinian doctrine first published in its entirety in 1859 in *The Origin of Species*. Darwin refined his mass of observations to a simple key which explained the mechanism of evolution in man, plants and animals in one bold stroke. Sir Julian Huxley has provided a useful summary of Darwin's observed facts and the deductions from them. These are:

1. The tendency for the numbers of all organisms to increase geometrically.
2. yet numbers are usually constant:
   (*a*) therefore there is a struggle for existence
3. variation between individuals, some being better adapted than others:
   (*b*) therefore there will be natural selection:
4. some of this variation is inherited:
   (*c*) therefore the effects of differential survival accumulate, i.e. there will be permanent change in the populations.

The first deduction was the struggle for existence 'beneath the peaceful face of nature' and the second, the natural selection of useful variations in that struggle—'even the smallest grain in the balance, in the long run, must tell on which death must fall and which shall survive'. Although Darwin emphasized the struggle for existence, actual struggle or combat is rare. One human or animal variety could replace another if it was more fertile and the fertility was inherited. But among animals and plants there *is* intense competition for available supplies of food and so only a few can survive to maturity. These, as Darwin said, are 'selected by nature' to contribute more to the ancestry of future generations. The less adapted forms will tend to die out and consequently will appear less frequently in the ancestry of future generations. The great strength of Darwin's theory is that it depends upon the working of processes that can be observed and is not merely theoretical. Darwin knew nothing of the blind and random process of mutation or the automatic mechanism of shuffling the genes which produce all the variety that he saw and which natural selection channels purposefully into adaptations. Our knowledge of mutations and the mechanism of recombination came after Darwin's life but they have done nothing to decrease the truth of his 'selection' theory. Rather they have strengthened it.

Can evolution by natural selection be observed now, going on around us? The answer is yes! To take a common example, the peppered moth has two varieties: a grey form which is concealed against lichen-encrusted stones and bark but is easily seen against a background of soot; a black form, which is produced by a single dominant mutation and is well camouflaged against sooty stones and bark in industrial districts but is conspicuous against lichens. A century ago the black variety formed less than 1% of the population of peppered moths in Manchester. Now in such industrial districts it forms 99% of the population. In clean areas however the grey form is still dominant although as industrial pollution spreads into the country so does the black form of moth. And selection has made the black form blacker than when it was first noticed. The selective agents are birds who spot and eat the conspicuous grey form but miss most of the black moths in cities. Indeed, here the black form survives over a 24-hour period 10% better than the grey form but 17% less well in unpolluted areas. Not only is the black form on the increase because it is less easily seen by birds against industrial grime but it is also tougher than the grey form. It resists cold better and the larvae can survive on polluted food. This single example, and there are many others, shows that selection *does* happen and that the members of a population, in this case moths, do make unequal contributions to the ancestry of future generations.

HUMAN EVOLUTION

But what about human evolution? Does variation occur and if so is it adaptive? Can natural selection be shown to be at work? Certainly the genetic variability of man is far greater than that of any wild animal. This is probably due mainly to the human species having begun, perhaps half a million years ago, to differentiate into separate races which, by migration or conquest, followed by intermarriage, has led to enormous variation. When the races started to diverge man, like any other animal, became adapted to regional differences in climate, disease and way of life. They diverged most noticeably in skin colour, skull shape and organs of speech (tongue shape and tongue movement giving rise to different languages). In spite of mixing, and even though differences are only now moderate, it is clear, as the chapters on 'race' show, that whites, blacks and yellows still differ somewhat in their physical genetics, their susceptibility and adaptation to disease and climate, and in their temperament. 'Genetic group difference', as Sir Julian Huxley has called it, still exists and warrants most careful study.

Can selection be seen at work in man? Infective disease is still one of the most powerful agents of natural selection in man; its effects can be shown by sickle-cell gene frequency occurring in populations exposed to malaria or free from it (see Fig. 14). Put simply (Chapter 12 describes its role in greater detail) the sickle-cell gene confers on the carrier of *one* gene (i.e. the heterozygote) the ability to resist the malarial parasite in one of two ways. One is through the production of haemoglobins which the parasite cannot digest. The other happens when the abnormal haemoglobin causes an infected red cell to sickle (i.e. to become sickle-shaped) because of the uptake of oxygen by the parasite; the sickle cell is then destroyed with its parasite. A double dose of the gene causes severe anaemia and is fatal to the carrier in infancy. Such a gene as this not only protects against malaria but it used to protect the human populations possessing it from disturbance by other less well adapted peoples. One gene therefore helped in the past to maintain the 'purity' of many tropical populations from invasion by people not adapted to resist malaria. The control of malaria has, of course, made the gene redundant in many areas, and as a result populations will become fitter. This is because the single gene which protects the heterozygote from malaria also causes a lowered state of health in its carrier, probably because the sickling reduces the amount of oxygen in the blood.

PROTECTION AGAINST MALARIA BY BLOOD TYPES

WORLD DISTRIBUTION OF MAJOR HAEMOGLOBIN ABNORMALITIES

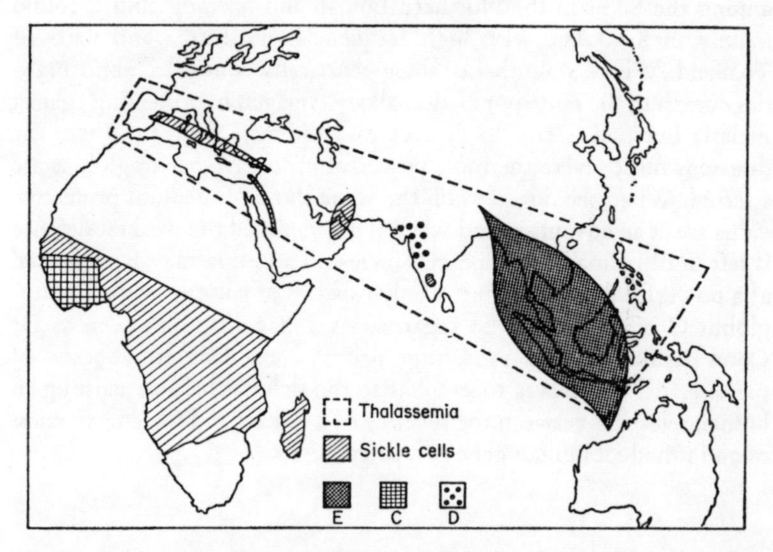

FIG. 14 Sickle cell anaemia has been mentioned in the text. Thalassemia can occur in two forms, major and minor. The major form is the result of a double dose of the thalassemia gene (Th) and is often fatal because of a severe anaemia and under-production of haemoglobin. The minor form is very variable. Sometimes it is difficult to distinguish it from a case of major thalassemia, but the minor form is produced by a single dose of the gene. If its partner gene is responsible for producing normal haemoglobin, the disease is mild. If its partner produces C or E type the symptoms are worse. There is no known protective effect of the thalassemia gene (or the genes responsible for producing other haemoglobin types, except sickle cell) but high frequencies tend to occur in places where malignant tertian malaria was endemic until recent times. A double dose of the C, D and E type of haemoglobin produces only mild anaemia It will be seen on the map that C type has a high frequency in Ghana and adjacent territories, D has a high frequency (2%) among the Sikhs of the Punjab while the main focus for the E type is in S. E. Asia (20% in Burma and parts of Thailand). It is likely that some selective advantage will be found in individuals possessing the genes for abnormal haemoglobins and it is likely that they will be found to resist the different forms of malarial parasites.

After Stansbury *et al.*, *The Metabolic Basis of Inherited Disease*, (McGraw Hill, 1960).

Various other abnormal haemoglobins besides that causing sickle-cell anaemia are foes to malaria. Some of these are: haemoglobin C, local to Northern Ghana and adjacent regions; haemoglobin D found among the Sikhs in the Northern Punjab and haemoglobin E found mainly in S.E. Asia with high frequencies in Burma and parts of Thailand. Where a number of these genetically controlled abnormalities coexist in the same population all are engaged in protecting against malaria but each seems to protect to a different degree. Where the disease is most severe the most powerful protective haemoglobins are selected. When the disease is on the wane those of medium protective value are at an advantage and when it has vanished the weakest defence is left until it finally disappears. Sickle-cell anaemia may be regarded as a powerful defence against a lethal species of parasite while haemoglobin C may possibly be regarded as a 'tapering-off' gene as Dr Coon has described it, protecting against a less dangerous species of parasite. All this serves to emphasize the delicate balance existing in human selection between the severity of a disease and the effectiveness of an individual human gene as a defence.

BLOOD GROUPS AND DISEASE

Members of the different blood groups in man are probably unequally susceptible to different diseases. Among more than 7,000 sufferers from duodenal ulcers in Britain, 55% belonged to the O blood group. In a control group (not affected by the disease) only 47% had an O blood group. Another disadvantage of belonging to the O blood group is that O individuals are also slightly more likely to suffer from gastric ulcers than individuals with other blood groups. On the other hand possessors of the O group are less liable to cancer of the stomach, sugar diabetes, and pernicious anaemia, diseases to which members of the A group are more prone. Table 8 summarizes blood group associations with some disorders of the food canal. The genes controlling blood groups therefore are not of neutral survival value; each seems to have advantages and disadvantages. Even so their meaning in relation to natural selection is very hard to interpret in modern society. Cancer of the stomach generally affects people past the age of reproduction and would not therefore affect, very much, the proportions of genes left in successive generations. More likely to affect this are diseases like toxaemia in pregnancy or broncho-pneumonia during infancy to

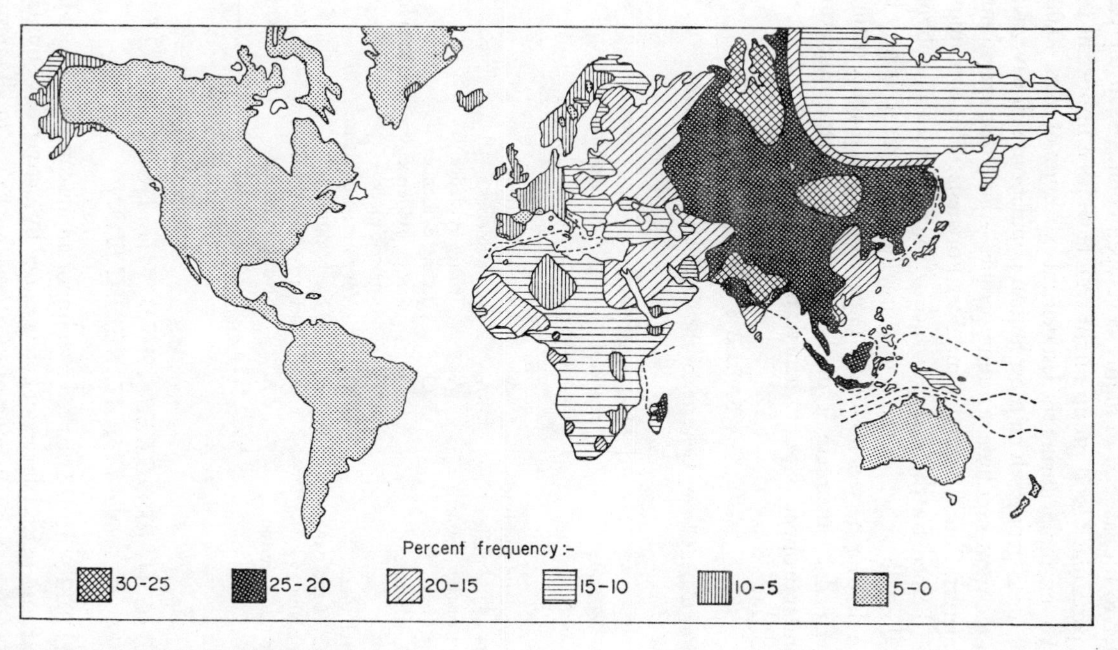

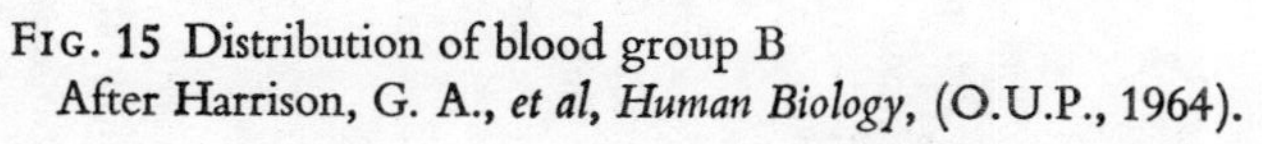

FIG. 15 Distribution of blood group B
After Harrison, G. A., *et al*, *Human Biology*, (O.U.P., 1964).

which carriers of the genes controlling the O and A blood groups respectively are prone. But in these cases medical advance has reduced real danger to life, at least in highly developed countries. In the past having a particular blood group might well have been important for survival. For example, during the last general expansion of man, 40,000 years ago, the B group which most human populations still have, was lost from the most rapidly expanding peoples—those that spread into South America, South Australia and East Polynesia. This loss might have been helped by the disappearance of such diseases as bubonic plague and smallpox which the B gene might help to resist. It may be that, since the B gene was no longer required to resist plague, natural selection failed to maintain it in the population of the migrating peoples and it dropped out of its genetic make up (see Fig. 15).

TABLE 8

After C. A. Clarke. Genetics for the Clinician. *Blackwell 1962.*

Condition	*Blood group associated with disease*	*Remarks*
Salivary gland tumours	A	Strong evidence
Cancer of the gullet	A	No evidence
Cancer of the stomach	A	Strong evidence
Gastric ulcer	O	Strong evidence
Duodenal ulcer	O	Strong evidence, stronger than in gastric ulcer
Stomach ulcer	O	Very strong evidence
Cancer of the pancreas	A	Fairly significant
Sugar diabetes	A	Fairly strong evidence
Pernicious anaemia	A	Strong evidence

MEDICAL PROGRESS AND THE DECLINE OF NATURAL SELECTION

The years 1912 and 1922 marked two important discoveries; a cure for pyloric stenosis (a fatal disease of babies caused by an over-development of muscle fibres at the opening of the stomach into the small intestine which blocks the exit of food) and the discovery of insulin. Since this last discovery the life expectancy and health of diabetics have increased and they are now able to have children more successfully. Both these defects are common but, as a result of surgery and insulin treatment, the genes controlling them must be spreading through the population (see Fig. 16 for pyloric stenosis). Harelip and cleft palate are also relatively common; modern surgery has helped to keep alive babies with cleft palates who previously might have died from feeding difficulties or respiratory infections. It has also helped to improve the looks of sufferers from these diseases and thereby assisted their chances of marriage.

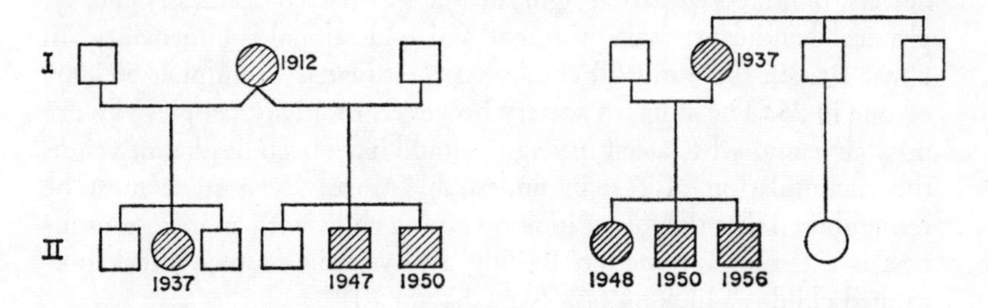

FIG. 16 Two family trees of pyloric stenosis; mother and children affected. The mother in the first family tree, born in 1912, was one of the few survivors before the operation to cure the abnormality was perfected (1912). All the other cases in both trees were successfully treated. Medical advance has therefore allowed people with this genetically determined disorder to survive and reproduce thus increasing the load of abnormal genes in the population. In fact about 8% of the brothers or sons of affected men are affected and no less than about 20% of the brothers and sons of affected women. It can be predicted that all medically and surgically curable genetic defects will increase in populations enjoying good medical and surgical treatment. It is a dilemma which confronts society now.

Pedigrees after Carter, C. O., *Human Heredity* (Pelican Books 1962).

These four common defects are inherited and the mutation rate from the normal to the abnormal genes controlling them is unchanged. Medical advance has allowed sufferers from these diseases to reproduce and pass on the traits to their children. This, together with the unchanged mutation rate bringing fresh diseases will lead to a higher accumulation of abnormal genes in a population. Indeed the table shows that genetic diseases have mounted in relative frequency while the curable infectious diseases have declined dramatically over 40 years (see Table 9).

There are many other rarer defects too which can now be cured and all these add up to a considerable load of 'bad' genes in a population. Of course the control of the infectious and contagious diseases of childhood together with better food and living conditions have allowed thousands to live and reproduce who would have died in childhood. A recent survey* has pointed out that hundreds of children who would have died in the past of infectious disease have been saved but often with long-lasting or chronic defects. The physical and mental handicaps which are taking the place of infectious disease are: speech defects, blindness or partial sight, deafness or partial deafness, epilepsy, physical handicaps, maladjustment and educational subnormality. In Great Britain these invalid children are estimated to number 500,000 or one in 25. The value to society however, of many people who are now alive and who, a century ago, would have been dead, outweighs the accumulation of certain undesirable traits. Even so it must be remembered that the job of looking after a child with an infectious illness is a temporary matter. Raising a physically or mentally handicapped child is a lifelong task (see Chapter 10).

TABLE 9

Percentage causes of death in the Hospital for Sick Children, London, from Genetics and Man, *C. D. Darlington, 1964, after C. O. Carter, 1956.*

Year	*Infectious**	*Non-Infectious*	
		External malformation	*Internal† errors*
1914	68·0	15·5	16·5
1934	51·5	32·5	16·0
1954	14·5	48·0	37·5

* Including T.B., Pneumonia, Gastro-Enteritis.

† Including pyloric stenosis and sugar diabetes (diseases with a known genetic component).

SOCIAL EVOLUTION

Man is quite different from all other animals in having two kinds of heredity, one of which in common with other animals, is based on

* *Today's Invalid Child*—a brief survey—see Bibliography.

chromosomes and genes. The other is a different kind of heredity which is based on the human brain. Man not only learns, stores information, and relates past experiences with present information to make decisions: he can hand on his information, his skills and his experience from one to another or from one generation to another in such forms as books, films, works of art, inventions, and of course, through speech. This transmission of ideas and fruits of ideas is what Sir Julian Huxley has called 'social evolution'. This social inheritance of knowledge and its results are fast, while biological evolution, based on genes and chromosomes, is slow. It took man perhaps 20 million years to evolve biologically from some ape-like creature (probably not like living apes). Yet only as recently as 40,000 years ago did man's skill begin to spread and show itself in the variety of tools he could make for hunting, fishing, boat-building, basketwork and much else. These skills allowed him to gather more food and extend his range; they were passed on to others by social evolution and probably favoured (by natural selection) greater elaboration of speech and its control.

More recently, in 8000 B.C. in South-west Asia and Central America, and independently in 5000 B.C. in Egypt and Palestine, man observed that edible seeds grew into plants that produced more seeds; hence by his intelligence and co-operative skill he discovered how to propagate wheat and barley which previously had only been gathered and stored for eating. These were the foundations of agriculture and the know-how was passed on from generation to generation. The new crops depended on man for their propagation and health and man depended upon the crops for his food and fertility, in fact for his survival; 'man and crop', therefore, became a new, closely related unit in evolution. Those men who did not improve in their skill were eliminated. And the skills that were at a premium were based on temperament—patience, foresight, industry and stability were all important for the grain cultivator. The wanderer and horseman on the other hand required a different temperament: adaptability, a capacity to live off the land, a love of animals, but lacked the power to control, the determination to carry on.

Metal working started about 3000 B.C. and the skill was passed on from father to son. An inventive skill was needed and successful (inventive) families tended to marry into successful families, increased in proportion to their success and migrated wherever they found markets. The inventive processes, therefore, involved a *disruptive selection* of the inventors from their kinsmen. The descendants of these people are still found as metal working castes in Arabia, India and Africa. The Negro blacksmiths who live in the Sahara are an example of a successful caste. They not only work iron but act as advisers to tribal chiefs and as couriers between them. Without high intelligence, no doubt, preserved by inbreeding, they would not be able to do their many jobs.

These few examples show that social evolution is relatively much quicker than biological evolution though it is not easy to disentangle the two. Professor Darlington has reminded us that those who say that human evolution is primarily cultural and only *secondarily* genetic forget that the chromosomes in the cell determine to a large extent what we can learn, what we can create and enjoy and how we behave. The skills and temperaments of the peoples described were no doubt genetically determined and they chose instinctively the way of life that suited them.

Social inheritance in man makes it possible for one individual to influence millions of people in his own or later generations. Winston Churchill was such a man, Hitler another (though warped) genius. Leonardo da Vinci, Plato, Mahomet, Lenin, Darwin, Newton were others, and there are countless more. Konrad Lorenz neatly expressed the effects of social evolution by citing the example of Faraday. If Faraday's behaviour when he built the first electric motor had been determined by an electric-motor-building gene we should not now be enjoying the benefits of an electromagnetic culture. Faraday had no children to perpetuate the gene and even if he had, its spread through the population would have taken thousands of years. In animal evolution, however, it is the average that counts while the unique are often quickly eliminated. But thanks to man's speech and accumulated wisdom and experience, the exceptional individual can and has influenced the course of human history.

10: Eugenics

> Science finds out ingenious ways to kill
> Strong men, and keep alive the weak and ill,
> That these a sickly progeny may breed;
> Too poor to tax, too numerous to feed.

DO we know enough yet about human genetics to improve the human stock—to breed for health, adaptability (important in a society where automation will continually replace men), intelligence, vitality and emotional stability? The answer is NO! Mendel's work was rediscovered only 65 years ago and it is not surprising that we are not even on the fringes of improving, say, intelligence. One fundamental difficulty is to know what to breed for. Emotional stability is one trait that is desirable but although its presence in everyone would remove the maladjusted it is these who often enrich civilization by their genius. Again, to try to breed a super-race of highly intelligent men and women has its dangers, recognized by a great geneticist William Bateson, who said, 'I would trust Shakespeare, but I would not trust a committee of Shakespeares'. The eugenist knows however that the improvement of the hereditary qualities of the human race rests on the basis of the laws of heredity. Thus, he knows that there is an enormous range of genetic variation in desirable qualities. He would not wish to aim at improving intelligence at the expense of health or artistic ability. Nevertheless, genetics has shown that many people are cruelly handicapped by their genes and that the whole of the human species suffers from this evolutionary burden. Perhaps in a century it may be possible to detect the 50 or more recessive genes in carriers, by urine, blood and other tests, while an examination for these genes may form a crucial part of the health service. But then what? How can scientific eugenics get rid of these 'bad' genes, two or three of which are perhaps carried by everyone without their knowledge? In the forseeable future it seems an impossible task. To take one concrete example to show the difficulties, that of cystic fibrosis. This is a disease of the pancreas which kills one baby in every 2,000 born in Britain and probably 4% of the people in this country carry the recessive gene that causes it in double dose. (Fig. 17 shows that this recessive gene may be detected in the carriers). This means that about two million people would have to be stopped from having children to get rid of

DETECTION OF CYSTIC FIBROSIS OF THE PANCREAS

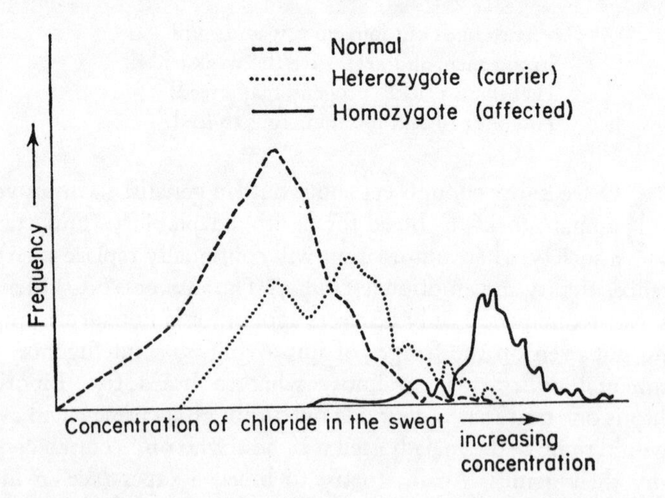

FIG. 17 It will be noticed that on the whole the concentration of chlorides in the sweat of carriers is midway between that of the normal and affected homozygotes, but all three distributions overlap greatly.

After Stern, C., *Principles of Human Genetics* (Freeman 1960).

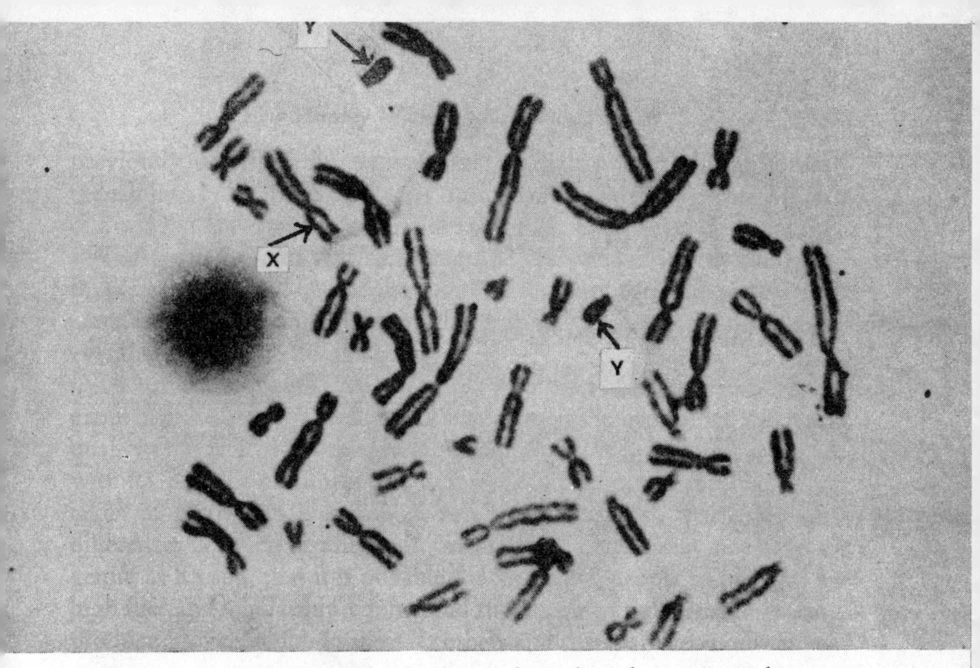

PLATE 3—(*a*) *above*. Cell from a blood culture of a male with an extra Y chromosome as it appears under the microscope. (*b*) *below*. The same XYY cell as in 3*a* but with the chromosomes arranged according to the Denver classification. The chromosomes have been cut out of a photograph and matched in pairs. *By permission of Dr M. D. Casey*.

1 2 3 4–5

6 — 12 + X

13 — 15 16 17—18 19—20

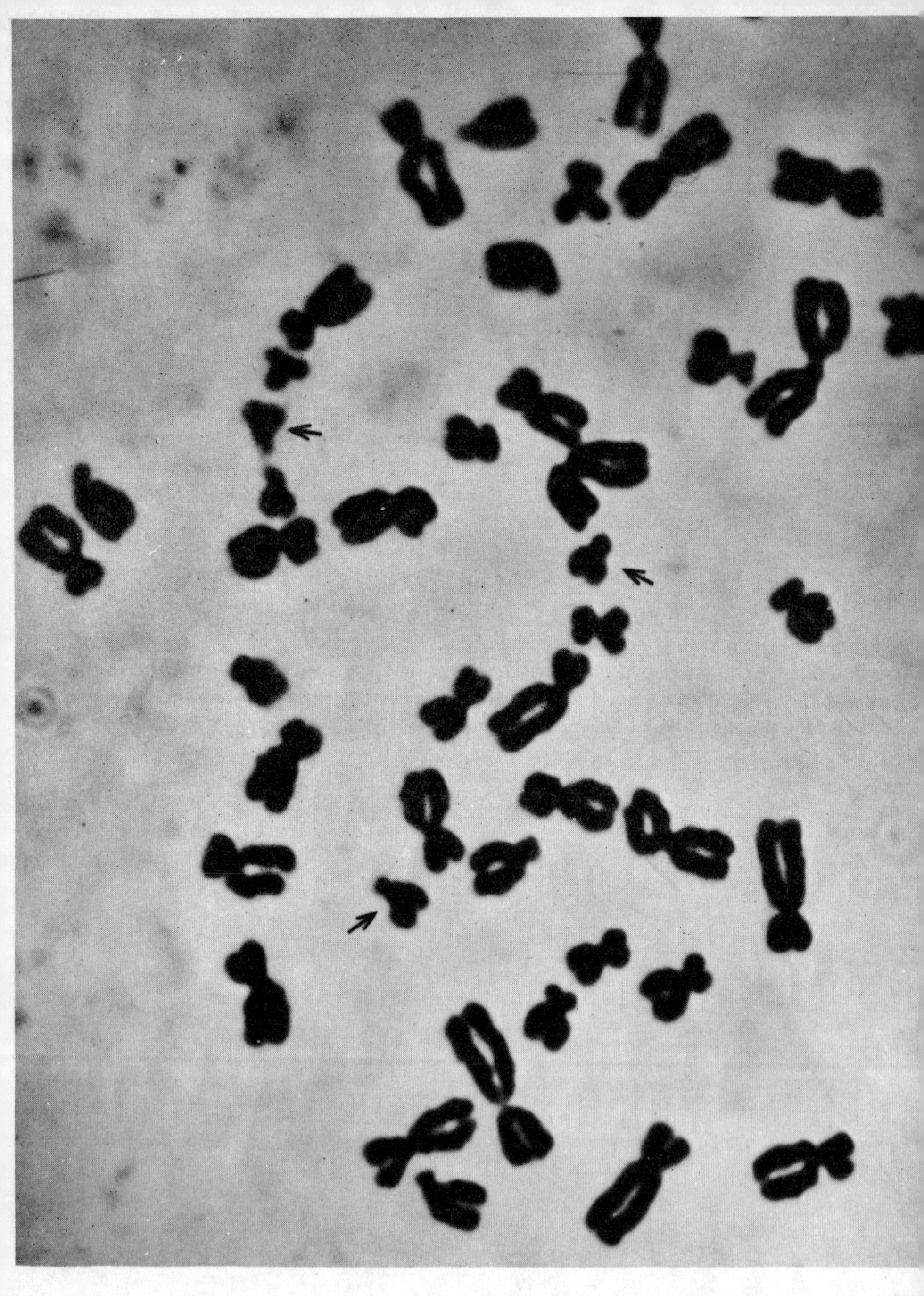

PLATE 4—The mitotic complement of a *mongoloid* human female showing 47 chromosomes. × 5,500. Chromosome 21 (arrowed) is present in triplicate.

Preparation by courtesy of Dr Peter Pearson. Photo by Dr A. Haque and Mr C. G. Vosa, Botany School, Oxford.

the gene from the population—clearly an impossibility. But a married couple with one diseased child should be told that one in four of any other children is likely to die of the disease. Brothers and sisters of similar children should be warned against marrying first cousins where there is a 1 in 24 chance of having a diseased child; if they marry an unrelated person the chance is 1 in 135. Indeed in any typical recessive disease, carriers, if they can be detected, should be discouraged from marrying each other by telling them what is likely to happen if they do. If this policy could be carried out the overt incidence of diseases like fibrocystic disease of the pancreas and phenylketonuria would fall almost to zero between one generation and the next.

Another example is given in Table 10 which summarizes the risks of having a child with harelip, with or without a cleft palate. All this advice might be called *negative* eugenics for it is designed to reduce the birth rate among people with defective heredity.

More drastic measures than advice have been advocated for the

TABLE 10

From Genetics for the Clinician *by C. A. Clarke, Blackwell, 1962.*

Cleft palate	*Harelip (with or without cleft palate)*
If an affected child is born to normal parents and there is no affected close relative* the chances against a further child being affected are about 80 to 1	If an affected child is born to normal parents the chances against a further affected child are about 20 to 1
If an affected child is born to normal parents and there is also an affected close relative, the chances against a further child being affected are 10 to 1	If one parent is affected the chances against any child being affected are 50 to 1
	If one parent is affected and there is also an affected child the risk to subsequent children is about 1 in 10

* Close relative means parents, children, brothers and sisters.

insane and feeble-minded. Sterilization has been used for over sixty years in the United States and for more than thirty in Scandinavia. In Copenhagen over 8,000 legalized abortions and over 1,100 legalized sterlizations have been made in the past twenty years. These measures would certainly cut down the number of mentally defective people born to defectives but the fact remains that most defectives come from marriages between normal people. Even so, genetic counselling might help to spot in family trees a defect which might appear in the children of a marriage, so that couples are well aware of possible consequences and can act accordingly.

Positive eugenics, that is a planned programme to make the *population* (not a lineage or 'family tree') as a whole more energetic, less prone to disease and more talented, has not met with as much success as negative eugenics for the reasons stated at the beginning of this chapter. Another is that such a programme would interfere intolerably with personal liberty. It would mean for instance that any one woman would be strictly limited in the choice of a father for her children and *vice versa*. Clearly this is outside the bounds of all possibility, even if it were right.

One possible positive way of propagating intelligence or some other quality (see Chapter 7 for a discussion on the alleged decline of intelligence) might be the use of artificial insemination by the sperm of a donor of a couple's choice (A.I.D.) to remedy a sterile marriage. With the setting up of sperm banks, and perhaps in time, egg banks, it might be possible to give a population a new spectrum of genotypes. But here there is a serious qualification: if the results of voluntary 'germinal choice' (A.I.D.) are kept secret there will never be any possibility of assessing, correcting, or improving on them.

More far-fetched than A.I.D., which has been going on quietly for years, are the startling eugenic proposals made by Joshua Lederberg, one of which is to increase the grey matter of foetal brains by injecting them with growth hormone before the number of nerve cells has been fixed. Another is to use viruses to transfer D.N.A., extracted from, say, the somatic cells of a genius, into human reproductive cells thereby altering the person's genes. A third is to use laser beams to slice up the chromosomes of reproductive cells, cultured *in vitro*, and combine certain pieces to 'engineer' a human being. The late J. B. S. Haldane in a highly speculative moment suggested that such chromosome grafts might eventually be used to make men suited for extra-terrestrial exploration—'prehensile feet, no appreciable heels and an ape-like pelvis' would be useful adaptations for the crew of the first space ship to the Alpha Centauri system. Their weight would be reduced and also their food and oxygen requirements.

The present state of affairs is, however, that there is little to be done in the way of negative eugenics other than persuading certain people not to have children. The knowledge necessary to operate a programme of positive eugenics involving the genetic engineering mentioned above might well take a thousand years to accumulate. In any case people will have to consent to any plan. Certain basic steps have been suggested, however, to educate public opinion. For one thing people might be made aware that less intelligent people tend to propagate their like faster than the more intelligent, leading not only to a quantitative population crisis but a *qualitative* one too. Presentation of such facts might lead to the more intelligent having more children but this is doubtful unless special bonuses are awarded or tax relief allowed for each child above a certain number. Such a plan would be highly controversial for it would mean that certain groups of people would be singled out for special financial treatment. In short, there is nothing much that can be done except by a slow process of educating the public in the sciences of genetics and eugenics and where appropriate using sperm from suitable donors. If this is done the public can then make their own decisions in the light of the facts.

Part 2: Race

In many ways the major races of man are like the sub-species of any other mammal. They form groups whose members resemble each other by having their own average combination of physical characteristics yet who differ from other groups of the species in certain persistent (and sometimes vague) ways. In animals breeding is possible between sub-species, hence there is no sharp boundary of characters between them. So it is with the races or sub-species of man. In the long past huge distances, seas and mountain barriers led to the formation of these sub-species each adapted genetically to its own climate, food, altitude and local disease. Conquest and migration, followed by crossing in the recent past, has caused each human sub-species to shade one into the other so that although the 'typical' Negro, Chinese and Englishman belong to different peoples each is connected to the other by every gradation of type. If each sub-species had continued to evolve in isolation, on the basis of the evidence from animal evolution, it would have become a species in its own right, unable to form normal, healthy offspring with other species. We know, however, that although there are still considerable differences between people in their looks and in their genes, cross-marriages are normally fertile and the children are healthy and intelligent. Although therefore there may be a pile-up of genes in one place to give, for example, a brown skin here and a pink skin there, human genes are common property. That is they can be shuffled from people to people by marriage so that the human *species is able to go on evolving as a unit.*

In the next few hundred years race mixture facilitated by ease of transport may have gone so far that peoples like Bushmen, Eskimos, American Indians, Sherpas and Australian Aborigines, at present not touched greatly by technological civilization, will have been merged into a world community. We know very little about how such peoples are adapted to climate, altitude and disease and we are now presented with probably the last chance to study them before their special characteristics are broken up by mixture.

The tools of statistics and genetics, as we shall see, have demolished any fact of racial superiority; they cannot demolish as rapidly, however, superstition, prejudice, false argument and discrimination.

11: Race (1) Pinks, Blacks, Browns and Yellows

IF a man from Mars landed on the earth and collected a 'typical' Englishman, Negro and Chinese for detailed study on his planet he might at first think that they belonged to three different species. Skin colour, nose, lips, and shape of head would be different and so would colour, hair form, shape of eye and its colour, and body build. If this collection interested him and he wished to make further observations by travelling round the world his first clear-cut ideas would certainly change and become confused, and eventually he would have to say that clear-cut divisions between different races just do not exist. However different man may seem at the extremes, say in China or West Africa, or Sweden, his racial distinctions shade into one another and are lost through intermarriage. For example, in Western Europe a number of races have been described; the Nordics with fair hair, blue eyes, large build and long heads; the Alpines, characterized by medium hair, brown or blue eyes and very round heads, and the Mediterraneans, by dark hair and eyes and long heads but smaller in stature than the Nordics. Even in Sweden however, where a high proportion of the Nordic type would be expected, only about 11% were found among a batch of army recruits; the rest were Alpine and Mediterranean types and mixtures of all three. Anyone walking down a street in, say, Derby, would see all three types and their intermediates. This mixing on a small scale is reflected, on a larger scale, in vast areas of intergrading like Africa just south of the Sahara or the melting-pot of the Amazon Basin, where short, flat-faced Mongolian types rub shoulders with people with unusually wavy hair and faces rather like those of dark Europeans. The important point is that man is a vast world-wide species which is very variable in any one place yet the *average* individual from any one country also varies from place to place. If, for example, we went out into the street and picked haphazardly a thousand Englishmen of European origin and compared them with a thousand similarly chosen Ghanaians of West African origin for skin colour, hair characters, skull shape, stature and intelligence we should almost certainly find that even the darkest Englishman would be lighter than the lightest Ghanian and probably the curliest-haired Englishman would be straighter-haired than the straightest-haired Negro; but in stature, skull shape and intelligence there would be a great deal of similarity. If indeed, our two thousand samples were skinned and their faces masked it would be extremely difficult to separate the two groups. But although there will be a lot of variation among the Negroes and the Englishmen, the 'average' Negro and 'average' Englishman are obviously different to look at and are part of the four great racial stocks, the white, the yellow, the black, and the native Australians. The whites (or better, pinks) have the hairiest bodies, the thinnest lips, and the straightest faces. The Negroes have fuzzy hair, projecting faces and thick lips; the Mongoloids (yellows) have flat faces, opaque skins and the eye opening is narrowed while the native Australians are hairy, small-brained, big-browed, with protruding mouths and sloping foreheads and chins. In most peoples however, migration and crossing has blurred any sharp distinctions between the four stocks.

JEWISH AND GERMAN RACES

Many people claim to be able to pick out Jews by their looks, but when their claims are put to the test many mistakes are made in identification. Physically, the Jews are very variable and, due to marriage with non-Jews, there is a tendency for them to resemble the non-Jewish natives of a country. For example, Spanish Jews are Spanish in looks and a

Chinese Jew resembles other Chinese. Nevertheless, although there is no such thing as a Jewish race, Jews have certain characteristics which occur more frequently among Jews than among their non-Jewish neighbours. A few such characteristics may possibly be inborn and transmitted from generation to generation. Certainly their strong sense of religion, exclusive social system and sense of common suffering have held them together as a group whose members have tended to intermarry and so preserve their Eastern Mediterranean origins and, no doubt, their intelligence. But to consider the modern Jews as a race with extraordinary origins has no foundation.

The case for the existence of a German race, superior in brains and muscle, which was used for propaganda purposes before and during the Second World War, is even less convincing. The Germans are probably more variable, physically, than the English. In the North they are mostly of a Nordic type, fair haired and long headed. In the South they are mostly of the Alpine type, brown haired and round-headed. Nevertheless, no one would deny that Germany has a national spirit like most countries. It is this together with language and culture that forms the real basis of division in Europe and not such trivialities as the size of head, colour of hair or shape of nose.

BLOOD AND RACE

In 1956, at a meeting of the South African Medical Association, it was suggested that the blood from non-Europeans should not be used for transfusion to Europeans. A newspaper suggested that the transfusion of blood from one race to another should be stopped. Doctors agreed and disagreed. What is the layman to believe? Is the blood of one race better than another? Is there a difference in its structure? What differences are there between the blood of different races? Some eleven blood group systems are now known in man of which the ABO system and the Rhesus are the most familiar. Everyone can be classified according to the systems but will only have one particular combination within any one system.

To take one example, the ABO system; all people can be classified into four groups which can be A, B, AB or O. If the wrong blood is given to a patient during a transfusion the plasma of the patient's blood destroys the red cells of the donor's blood. The destroyed blood blocks the patient's kidney tubes and can cause death. It is thus extremely important to find out whether blood is of the right group before transfusion. Blood groups are permanent characteristics of an individual; climate, diet, illness or medical treatment cannot alter the group which is inherited from an individual's parents.

Could a scientist tell whether a spot of blood came from a Negro or from an Englishman? The answer is most probably no. But if tests were made on a thousand men and three hundred were found to belong to Group B, the population could not be English (see Table 11). Since the First World War (as a result of investigating the ABO system of the different populations making up the allied armies in Macedonia) it has been known that the frequencies of the ABO types in human populations differ from one population to another. Our unknown thousand could be Indian, Russian or Chinese and further blood tests could be made to decide which. The important point to understand is that most racial groups include all types of the ABO system: A, B, O or AB, but in differing proportions. That is they differ in the *incidence* of genes controlling blood types, not in having totally different sets of genes.

TABLE 11

Frequency (%) of blood groups

Group	*O*	*A*	*B*	*AB*
English	45·8	42·2	8·7	3·2
Russian	32·9	35·6	23·2	8·1
Chinese	45·5	22·6	25·0	6·1
Hindus	30·2	24·5	37·2	8·1

On the more local level, in North Wales the frequencies of blood groups of people with Welsh names differ slightly from those with non-Welsh names (see Table 12). There are more O and B individuals and fewer A and AB among the Joneses and Williamses than among people with English names. Obviously then the genetic make-up of the Welsh group is distinct from the English group among whom they live.

Brothers and sisters in the same family might easily have different blood types inherited from their parents, so that a brother's blood may be a greater menace in a transfusion to his brother or sister than that of an African bushman of the same blood group. There is no scientific

TABLE 12

Percentage frequencies of blood groups among blood donors with Welsh and non-Welsh family names.

Men and single women donors with	*Number of individuals*	*O*	*A*	*B*	*AB*
Welsh family names	909	52·7	35·0	9·7	2·6
Non-Welsh family names	1,091	46·6	42·0	8·3	3·2

basis for superiority about blood. The really important thing about blood transfusion is whether the blood is free from germs and can mix with the patient's blood without any ill-effects.

BLOOD GROUPS AS MARKERS

The blood group systems of a population sometimes provide useful markers of its ancestry since the blood groups are indestructable characters that are passed from generation to generation unchanged. For example, the ABO system of Hungarian gypsies, whose custom is to intermarry, is different from that of the rest of the Hungarian population and is almost identical with that of the people from north-west India from whom they came a thousand years ago. The Swiss Walsers who left the Valais in the thirteenth century and colonized Davos, Arosa, Avers and many other valleys in the Grisons, show a blood group pattern quite different from that of their neighbours and like that of the Valaisians whom they left seven hundred years ago.

The further East a population lives or has its origin, the higher is its percentage of the B blood group which must have originated in the East (see Fig. 15).

BLOOD GROUPS AND LANGUAGE

There may be a link too, between the ABO system and the development of language; to be more specific, the blood group gene O may affect popular preferences for pronunciations such as the English TH sound which also is present in the Spanish lisp and the Greek theta. This theory, proposed by Professor C. D. Darlington has not yet been accepted by anthropologists and many geneticists.

The TH sound, according to Darlington has been maintained on the edge of the Eurasian continent while it has vanished or failed to develop in the whole of the central area. Ability to pronounce this sound links Icelandic, Welsh, English, Basque, Castilian, Albanian, Greek and classical Hebrew and Arabic. Although the languages have an independent history they are spoken by peoples who are racially related since the majority of them (over 65%) possess the blood group gene O and are able to say the TH sound. Looking at the blood gene map of Europe (Fig. 18) it seems as if there has been an expansion of centrally placed peoples driving the O gene possessors to the edge.

Referring to the map, Europe can be divided into three O gene zones:

1. A peripheral zone from Iceland to Greece with 65 to 75%.
2. An intermediate zone from Sweden to France with 61·5 to 75%.
3. Two zones, a main eastern zone and the Portuguese corner, with 61% or less.

The first zone is the TH positive and unites Teutonic and Celtic in the north-west, Latin and Euskarian in the south-west, Greek and Albanian in the South. The second zone consists of all the countries which had TH but have now lost it. It survives only in the old spelling of personal and place-names. The third zone contains countries which have not had any recorded pronunciation of TH during the last 2,000 or 3,000 years.

What is interesting and important for us here is that different peoples to the southern and western edges of the Eurasian continent, the European, Arabic, the Dravidian with Burmo-Siamese (the Christian, Islamic and Buddhist) are linked by O blood group frequency and a feature of language. And this agreement of language of peoples is due to migration and conquest by centrally placed peoples over the last 4,000 years.

The development of language and dialect has not been sufficiently linked with differences in the speech apparatus of individuals, families, small communities and races, shape and size of tongue. Teeth, palate, mouth, throat and nasal cavities all affect speech and there is evidence that these variations are genetically determined and that they have affected the evolution of dialects and languages. The anatomical differences of the speech apparatus of the different races has received but meagre attention from anatomists, although William Lawrence in 1819

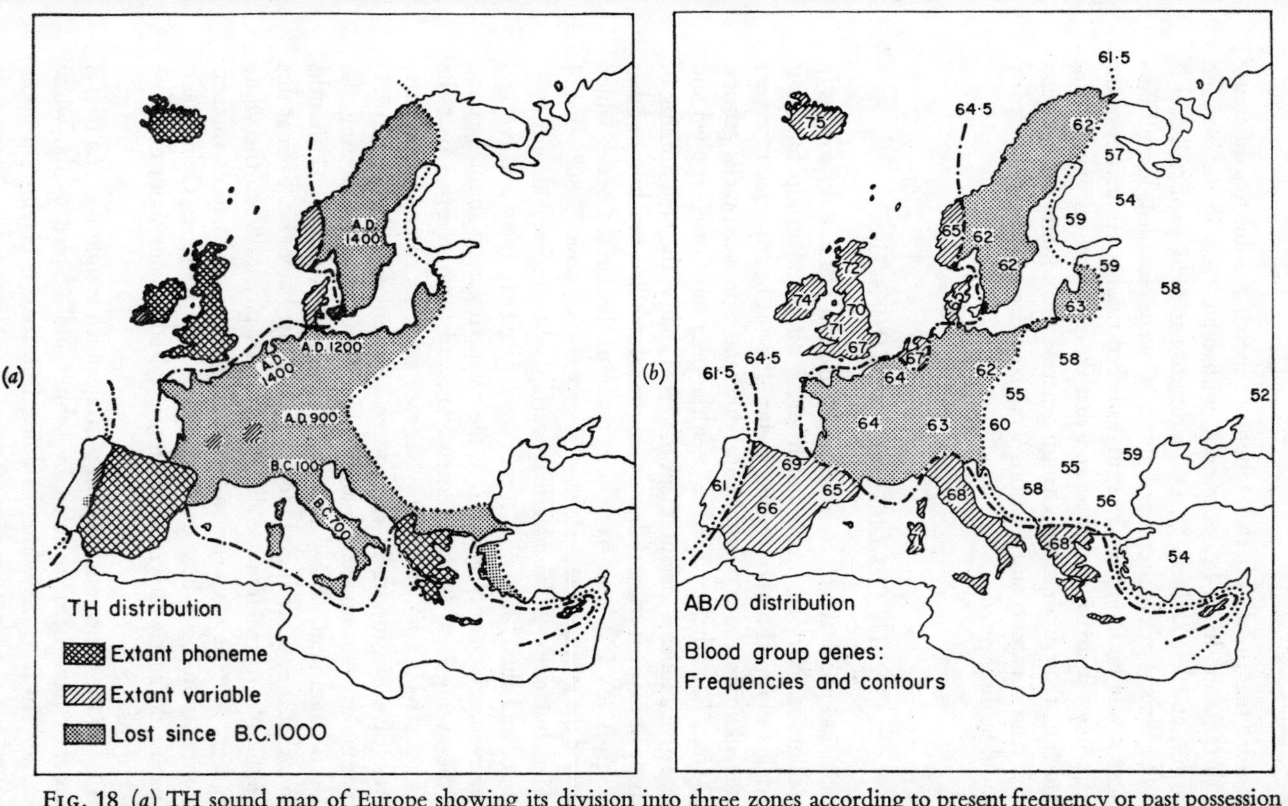

FIG. 18 (*a*) TH sound map of Europe showing its division into three zones according to present frequency or past possession of the θ and δ sounds. The dates in the middle zone mark the time of loss.

(*b*) Blood group map of Europe showing the regular gradients and a division into three zones by contour lines at arbitrary values of O-gene frequency in the population.

After Darlington, C. D., 'The Genetic Component of Language', *Heredity*, Vol. 1, Part 3, 1947.

recorded that the click in the Hottentot's speech was due to a difference in their palate shape. In comparison with other races it was 'smaller, shorter, and less arched'. C. D. Darlington in 1961 predicted that if speech defects of children aged about 10 were recorded, local differences would be found between undisturbed rural communities in the frequency of particular defects and that these differences would form gradients, their frequencies changing across the country in a regular way. And if parents and relations were recorded evidence of heredity would be discovered.

INTELLIGENCE AND RACE

Social and political factors bedevil the question of racial differences in intelligence. No one boasts much about his blood group but most people feel superior about being brainy. It must be admitted that very little is known about psychological differences between racial groups. I.Q. tests (see Chapter 7) have been the ones most often applied and these give a reliable measurement of verbal ability; they tell nothing of artistic or mathematical ability or emotional stability, or a host of other things. Many I.Q. tests have been made in the United States which showed the average intelligence of Negroes to be about 75–80% that of whites but there is considerable overlap when individuals are considered and about 25% of Negroes equal or surpass the white average. The average I.Q. of Negroes from the northern states is about the same as the average for the southern whites, and Ohio Negroes beat the Arkansas whites by 10 points. Negroes from Northern States score better than Negroes from the poorer, southern states so that an unfavourable environment certainly plays some part in lowering the Negro attainment. 'Baby' tests on Negro infants of 2 to 11 months show that they score slightly lower than white babies but even at this age the Negro children are slightly inferior, physically, to the white children, so that environment, including the months spent in their mothers' womb, had been poorer. In these American tests, Chinese and Japanese do nearly as well as whites, and Red Indians about the same as Negroes.

One of the enormous snags in putting much weight on such tests is that they have been invented by whites and adapted to the white situation. In fact, the white, in doing I.Q. tests, is playing on home ground and the Negro is playing away. Tests invented by Negroes or

Arabs applied to Englishmen, might give surprising results! Indeed in 1926 Helmer devised tests which contained subject-matter relating closely to the life of Red Indians. When white and Red Indian children did the tests the white children were definitely inferior. However it might be worth pressing on with the use of western type tests in developing countries like Africa because African aspirations are in the direction of a Western technological civilization. Another consideration worth pointing out is the difficulty of getting a fair sample to compare in a society where the members of races are differently treated. The conquered, or exploited race, will be at an economic disadvantage and we have seen that an unfavourable environment lowers performance in an I.Q. test. Moreover, if a section of the population is taken where economic conditions are similar a poor sample of the conquering race will be chosen.

The alleged superiority of whites over blacks therefore is not proven and even the present evidence of I.Q. tests shows, like most other characteristics, a good deal of overlap between white and black. Any large group—Chinese, Negroes, Europeans, will be found to overlap in intelligence and some of each group will have high and low 'intelligence'. If they do differ, the difference will be only a moderate one.

In Europe I.Q. tests on boys from Sicily, Flanders, Hanover, the Pyrenees and other places, showed very little difference and on this evidence no one could say that one 'race' is superior to another. Nearer home, this is what Francis Galton wrote about the Scots and English a century ago—

The average standard of the Lowland Scot, and the English North-country man is decidedly a fraction of a grade superior to that of the ordinary English because the number of the former who attain to eminence is far greater than the proportionate number of their race would have led us to expect. The same superiority is distinctly shown by a comparison of the well-being of the masses of the population; for the Scottish labourer is much less of a drudge than the Englishman of the Midland counties—he does his work better and 'lives his life' besides. The peasant women of Northumberland work all day in the fields and are not broken down by the work; on the contrary they take a pride in their effective labour as girls, and, when married, they attend well to the comfort of their homes. It is perfectly distressing to me to witness the draggled, drudged, mean look of the mass of individuals, especially of the women, that one meets in the streets of London and other purely English towns. The conditions of their life seem too hard for their constitutions, and to be crushing them into degeneracy.

This piece has been quoted because such differences are not obvious now. Transport and a better standard of living have ironed out the differences noticed by Galton and it is doubtful if the differences in intelligence he implied ever had any foundation.

12: Race (2) Biological Engineering

ADAPTATION OF RACES

IS an Englishman a better man than a Negro or Chinese in England and are the Negro and Chinese better men in their own countries? By 'better' is meant more resistant to disease and better adapted to their particular climate and habitat. Certain bodily characters do appear to be useful in certain conditions. For example, the people of two climatic extremes, the broiling deserts of the Nile and the frozen wastes round the Arctic ocean, appear to be well adapted to their climates. The lanky men herding cattle under the broiling sun along the White Nile in the Sudan are better able to avoid heatstroke by being long and thin and having a large surface of skin for body heat to escape through. The stocky, well-covered body of the Eskimo loses heat relatively slowly and is thus able to conserve important body heat in a cold climate. He has learned to run for long stretches behind his sledge, fast enough to keep him warm but not exhaust him. His hand can better withstand great cold, probably because of an increased blood flow, than can, say, that of an Englishman or Negro.

The face of an Eskimo, adapted to withstand frost, is a mask against the cold; its bones and flesh are modelled to give great protection against cold, dry air. The face is flat to reduce its surface area to frost-bite, and padded with fat to insulate against cold. The nose is fairly flat, the nostrils narrow to warm and moisten the cold, dry air and the high cheek bones help to protect the nose. The eyeball is protected by fat pads in the eye socket and the upper part of the socket is sheathed by a fold of fat-padded skin which reduces the eye opening to a slit to protect the eye against frost. Beards are scant so that the chances of moisture condensing round the mouth are reduced. This type of face was probably developed during the Ice Ages when the intense cold, pneumonia, and infected sinuses selected the peoples with this type of face as best adapted to survive in very cold climates. Since then the migration of peoples southwards brought this frost-adapted face to China, Japan, Tibet, tropical south-east Asia, and Indonesia.

Perhaps too the small stature, long arms and flexible muscles of the Pygmies are adaptations which are of real advantage in gliding easily through the dense forests they inhabit. The Pygmy smallness does not seem to be the result of poor food for their diet is good with plenty of protein and vegetables, but genetically determined. And the uneven storage of fat in the Hottentot and South African Bushmen causing their buttocks to protrude while the rest of their body remains thin may be adaptive. Hottentots and Bushmen are hunting, desert people and an all-over, heating blanket of fat may be disadvantageous to their conditions of life. In this respect they resemble the camel whose food store is similarly in one place, its hump. The buttocks are much more developed in the women and may be a useful fat reserve during pregnancy and lactation especially in conditions where food supplies are precarious.

But these are extremes. It is very difficult to try to concoct any explanation of, say, the bodily characteristics of the English as adapted to their rather damp, cold climate, (except that the nostrils are, like the Eskimos, narrow to warm cold air), although hidden adaptations to disease surely exist and are probably inherited. Indeed many variations between different peoples are of uncertain survival value. For example: most Chinese possess the gene for dry ear wax while most American whites do not; the eyeballs of the Negro are larger than those of the Japanese or the American white; the muscles controlling facial expression are different in Negroes and Italians while the finger print patterns of different peoples vary. Every difference between populations, such as the ones listed above, which is large enough to be significant and which has a genetic basis must have, or have had, a reason. Twenty years ago the ABO system was thought to have no survival value but we have seen that this is not so. In time evidence may well accumulate to show that some of its variations listed above are, or were, necessary for the conditions of life of the people who possess them.

Protection against malaria is certainly inherited and the presence of one inherited factor, or gene makes the blood corpuscles resistant to the attack of the malarial parasite. If, however, two genes are inherited, one from each parent (a double dose) death occurs from a form of anaemia, sickle-cell anaemia. A single dose of the gene is good for it protects against malaria, and children inheriting the single dose stand a better chance of surviving to reproduce. In some East African tribes 40% of the population has the sickle-cell gene but there is a good deal of variation. The incidence of the trait is usually between 20 and 30%. Does having this special gene cut off the Negroes as a race? No, because the same protective gene turns up wherever malaria is about. It is not uncommon in Greece, Italy, Turkey, Arabia and India. This is a good illustration of adaptation and also of another important point, namely that genes are not the special property of one race but can be passed to any. There are no limits. These inherited factors, or genes, tend to accumulate in a place or population giving rise to characters like a pink or dark skin, a frost-adapted face or resistance to a certain disease. This pile-up of genes is the real nature of race and the way in which populations begin to become distinct. Migration, followed by intermarriage between two differing populations spreads out the genes among them and racial distinctions begin to be broken up unless environmental conditions demand otherwise.

Perhaps a century ago adaptations to climate and disease resistance were more important than now though we have still much to learn about the adaptive relationships between heat, cold, altitude, and disease.

Medical advance has made the gene protecting against malaria redundant. Its frequency is now down to about 9% among North American Negroes whose ancestors left Africa about twelve generations ago. Of course we do not know the frequency of the sickle-cell gene among the original slaves, nor do we have much evidence on the amount of drop due to mixture with other populations. But if it is assumed that the slave population had a single dose (heterozygote) frequency of not less than 22%—based on the average in West Africa today, and allowing for one third admixture of populations, the heterozygote frequency should be about 15% in the United States today. Since it is only 9% it is reasonable to assume that the control of mosquitos has allowed the gene to drop out of the population.

A host of diseases besides malaria, which were killers fifty years ago, can now be cured so that West Africa is no longer the white man's grave. The nostalgia of the White man, Negro, or West Indian for home is probably of greater significance than physical illness, and can lead to mental illness, not easily cured.

Man, of whatever type, is highly adaptable and healthy men can quickly acclimatize themselves to heat and cold. Men on Artic expeditions often wear less clothing during the winter than at the beginning of their trip and acclimatization to heat is demonstrated continually in peoples of different origins living in hot climates. The Chinese have successfully survived in the thin air of Tibet. Whether they will be able to reproduce and maintain a population there for long is another matter. Our Olympic runners have certainly performed well in trials in Mexico City. In any case, in civilized countries we can now create our own climates in our homes so that a comfortable temperature can be maintained or the thickness of our clothing adjusted for bodily comfort.

MIXED BLOOD: RACE CROSSING

Many people are horrified by the idea of mixed marriage. In fact this, more than anything else, stirs up strong emotions and has provoked brutal hostility, not only in South Africa and America but nearer home, in Nottingham and Notting Hill. What are the biological facts about mixed marriages? Marriages between all the members of the human race can be fertile and their children normal, no matter whether the parents are Europeans, Africans, Americans, Indians, Chinese or Oceanics. Indeed race mixture has bred some of the most beautiful people in the world—the Hawaiians, the Balinese and the Thais.

It has already been pointed out that typical members of human populations differ from one another in looks, hair colour, skin colour, and in resistance to disease and climate and that these characteristics are due to gene outfits which have been built up over many thousands of years to fit a particular population to a particular place and climate. Is there any evidence to show how many genes races differ by? A little. For example the results of crosses between Bantu Negroids and Europeans indicate that the South Africans differ from the Europeans by only a few genes for colour and hair form, and marriages between Chinese and Europeans lead to the same conclusion. This statement is based on the fact that the children of the Europeans and white–Bantu half-castes often give children with white skins, blue eyes and straight, fair hair, while the children of Chinese–European marriages can have pink and white complexions, eyes without the characteristic slant, and brown hair. If larger numbers of genes separated the races, the physical characteristics would blend and would not separate out sharply. Crosses between West Africans and Europeans show this blending of characters in the children, indicating that larger numbers of genes are the cause of many of the physical differences between these particular populations.

Behaviour differences due to genetic recombination as a result of mixed marriage are illustrated by crosses between pure gypsies (rare now) and 'whites'. The offspring of the crosses are called, in England, the *Didikai*, in Germany, the *Jenischen*. The old, pure gypsy stock was artistic and musical and had elegance and wisdom. But they were wanderers because they somehow lacked the skill to invent and the determination to maintain, or the capability to organize, a life of settled cultivation. When they married, generally with the poorer type of white, many of whom were unsettled themselves, the old gene combinations were undone and recombined in new ways. The families produced were mixed in their temperaments, the temperament showing itself eventually by brothers and sisters marrying into opposed groups. Some changed the wandering life and became peasants, others remained as they were brought up. But the didikai is not so well adapted as the gypsy to a life so little supported by labour. The didikai goes in for peddling, rag and iron collecting, and sometimes petty crime. In general the didikai culture does not bear comparison with that of the pure gypsy. But smaller groups of the crossbred population have bettered their culture and economic status. These are the entertainers whose niche is the circus, the fair and the race-course, and whose society forms a hierachy of genetically varied and socially co-operating skills and abilities with the gypsy mixture in the lower layers, and the wealth, initiative and brains concentrated in the upper layers. This latter groups are probably crossbreeds who have lost the 'gypsy' genes of their ancestors by segregation.

When 'mixed marriage' takes place then, old combinations of genes are broken up and new or unusual ones are made. On a big scale this may lead to the creation of new cultures and skills such as the entertainers or may, on the other hand, destroy one old culture like the gypsies.

People having these new gene combinations are not at all unusual. Indeed, individuals with such gene mixtures may produce a variety in populations which is good. However, there is a natural brake to mixed marriage and hence to too much variety. This brake is the tendency for like to marry like and will preserve some of the extreme variations in human populations while intermarriage will provide novelty. Provided the brake continues to work the human race will continue to be interestingly divergent.

At present, then, there are no biological facts to support the theory that mixed marriages are harmful to the children of such marriages. The problem is mainly a psychological and social one. There are still many parts of the world where being a half-caste brings misery, especially to the black/white half-caste. The fact that mixed marriages involve special problems which affect the children of these marriages is due not to the mixing of genes but to colour prejudice and discrimination.

WHAT IS RACE?

Going back over the ground it might be useful to examine what 'race' means. It is a word that is often used loosely. For example, the Germans and Negroes are often described as musical races on the strength of one producing Bach, Beethoven, Wagner and Strauss and the other a string of famous modern trumpeters and jazz band leaders. The Italians are thought of as artistic, the Scots as mean, and so on, without considering that the average German or Negro may have little talent for music, while the Italian man in the street may have no feeling for painting or sculpture and your Scottish neighbours may be the most generous people you know. If race is defined, as it has been by some, as a group of people all of whom are alike in physical and mental characters, we immediately think of a set of identical people, quite devoid of that variability in shape, size, colour and so on that is present in every race. No one thinks of a race of red-heads, albinos or idiots; such a classification would cut through all the natural geographical groupings of people, for albinos, red-heads and idiots are found in every race. The above definition then, takes no account of variability or of the *geographical origin* of a people. A better definition is that given by the late Professor J. B. S. Haldane, who thought of race as a group whose members shared a certain set of inborn physical characters and a geographical origin within a certain area. So long as like tends to mate with like the members of these local varieties, or better, sub-species of man, will retain their own combinations of average physical and physiological characteristics. But in man with the present tendency for geographically separated populations to intermingle, and the increase in mixed marriage that accompanies intermingling, the grading of one sub-species into another is increasing and each generation will see fewer extreme racial types. Indeed in a few centuries they may have gone, so now is perhaps our last chance to study a 'baseline' from which man will evolve rapidly and perhaps develop new adaptations in response to urban pressures.

The sub-species of man as we have seen are very closely related indeed. Overlapping between them in physical and mental characteristics and in pronunciation are indications that relationship is very close. In the case of mental qualities we do not know for certain to what extent differences are due to nature or nurture or to a combination of both, but the evidence points to the former as being of great importance. We do know that all varieties of man can produce normal, healthy, intelligent children when they intermarry.

We have seen that some of the adaptations of the sub-species of man are useful, but others, like head shape or the thick, rolled-out lips of the Negro, are more difficult to explain as adaptations. They may possibly be accidents of evolution, although we should be careful in interpreting them in this way. The ABO system was thought to have no survival value twenty years ago but now it has been shown to be linked to susceptibility or resistance to certain diseases. And the anti-malaria genes show that certain peoples have adaptations to resist and survive attacks of local diseases. Perhaps other protective genes will be found too; useful, not only to protect against local diseases but to help to resist the rigours of climate and altitude.

Greater than any barrier set by inborn qualities of races are the barriers of language and culture and the even higher barrier of prejudice and discrimination.

Part 3: Man's Health and Food

Advances in medicine and better food and housing in rich countries like Britain have been responsible for the slow increase in man's life span from an average of about 40 years in the middle of the last century to about 70 now. Increased affluence, however, brings with it a crop of diseases which were uncommon even fifty years ago. This may be due partly to the lengthening life-span but not entirely. The developing countries present a different picture of health. Shortage of food, particularly protein, is responsible for a great deal of death and disease in infants. Because of chronic malnutrition they fall easy victims to infectious diseases that have been stamped out in Britain. Adults are weakened into a stupor by diseases such as malaria and sleeping sickness. The health problems of countries like Africa and India now must be very like the ones that faced England in the Industrial Revolution. But even in rich countries there is still a great deal of needless suffering from untreated and undetected mental and physical diseases. Some of these diseases are borne as part of life but some, like mild diabetes, the sufferers do not know about. We must have more preventive medicine so that undetected disease can be treated and cured and more health education so that ordinary people will know disease symptons needing medical attention. Doctors must learn to expect and seek out certain diseases in 'high risk' groups.

The food that people eat the world over has important bearings on health. It is worthy of note that taboos and religious convictions prevent good nutrition in many parts of the world while in others obesity prevails. Food sophistication can lead to obesity and this in turn to serious disease, since it generally provides sugary pap without the built-in restraint of bulk. Food improvers, food contaminants and residues left in food from agricultural practices need close attention from public analysts, but it is no overstatement to say that food here has never been so pure and that without pesticides there would be serious loss of crops to the detriment of mankind. Compared with the food foisted on the poor city dweller a century ago our food now is as pure as the driven snow.

What is the future of man's health? Is it possible for him to live to 200 fitted with a plastic heart? How artificial can man become? Are we, as J. B. S. Haldane said, too preoccupied with the sentence of death and not enough by the prison which is life?

ENJO

13: Diseases of the Affluent Society

A FEW years ago Professor John Yudkin showed that the increase in heart attacks (coronary thrombosis) could be neatly matched with the increase in motor-car licences, television sets and radios—all indicators of growing affluence. Heart attack was a rarity before 1921 when 743 men were certified as dying from it in England and Wales. In 1963 it killed about 100,000 people of whom 65,840 were men. Even though diagnosis has improved and has detected disease which previously went undetected, the increase is real and is partly due, of course, to the greater age which people live to nowadays. Not only has there been an increase in deaths from coronary thrombosis, but also in deaths from stroke, a degenerative disease of the arteries, lung cancer and bronchitis. The number of diabetics and people suffering from depressive illness has also risen. Tuberculosis and pneumonia, important killers 40 years ago, are not nearly so prevalent now. Some evidence for these statements is given in Fig. 19.

What are the causes of these new epidemics? Most of them are probably due to the way in which we live. The matching of the rise of coronary thrombosis with the increase in car licences points to lack of exercise as one cause of the disease. Certainly the effect of exercise

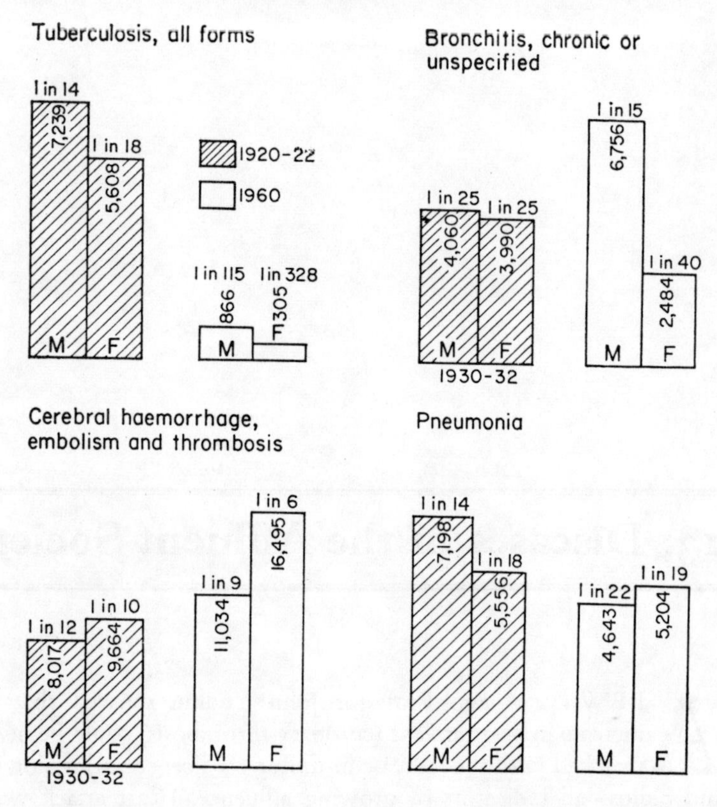

FIG. 19 Expected total deaths among 100,000 male and female persons (subject to the death rates for the year shown). The actual numbers are shown in the histograms.

After Backett, E. M., '*Towards Maintenance Medicine*', *New Society*, 16 July 1964.

might be to prevent youngish men from suffering a fatal heart attack. Drivers of double-decker buses die from thrombosis more often than bus conductors who have to run up and down the bus stairs all day. Office clerks and telephone operators suffer more than the walking or cycling postman. The lack of regular exercise is probably the only factor that doctors agree about as a cause of thrombosis. Indeed regular exercise such as walking is probably the best defence against ill-health of all kinds. It is possible that dietary habits may be another cause of the upsurge. In 1951 two Norwegian doctors pointed out that the death rate from heart attacks and strokes fell during the war in some

European countries soon after milk, butter, cheese and eggs were rationed. Deaths from these diseases continued to rise in countries with no shortage of these foods and once rationing was stopped death rates began to rise again. The culprit was thought to be cholesterol, a substance contained in certain meat fats, egg yolk and butter fat, which helps to fur up and thicken the arteries. This process is known to the layman as hardening of the arteries. Blood running through furred-up arteries seems to have a tendency to clot. When a clot occurs it is called a thrombus and if it happens to lodge in the coronary arteries it stops oxygen from getting to the working heart and this in turn can lead to sudden death.

Numerous surveys of the food habits of different peoples and communities have thrown light on a possible relationship between diet, cholesterol and thrombosis. In South Africa, for example, white, coloured and negro groups have a decreasing tendency for thrombosis which is paralleled by a declining fat consumption and decreasing cholesterol levels in the blood. Trappist monks are vegetarians, eat less fat and have a lower level of cholesterol in the blood and less tendency to thrombosis than do Benedictines who eat a normal diet.

Eating less animal fat is not the whole story however. Professor Yudkin has pointed out that in affluent societies, consumption of extracted sugar parallels fat intake. They both show a fourfold increase as between America and Canada, the richest countries, and India and Egypt, the poorest. Fat eating, according to Professor Yudkin, is not directly related to the increase in coronary thrombosis while eating sugar is. Increasing affluence and more efficient manufacture of biscuits, sweets, chocolates, cakes and soft drinks, have put up our sugar consumption to very high levels. During the past two centuries in Britain the annual sugar consumption has increased twenty-five-fold, from about 5 lb a head in 1760 to 120 lb in 1960. To test the idea that sugar is related to thrombosis, Yudkin showed that the sugar consumption of a group of 20 patients who had had a heart attack was 132 grams a day (mainly in cups of tea) while a group of 25 healthy men ate only 77 grams a day.

Fig. 20 shows the numbers of deaths from heart attack correlated with sugar consumption in different countries and broadly confirms that there is a definite relationship between sugar eating and coronary thrombosis. Even so it looks as if sugar is not the only cause; what have the French and the Dutch got that the Finns and the Americans haven't?

Besides dietary habits and lack of exercise other factors seem to be

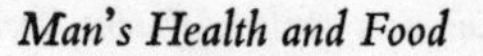

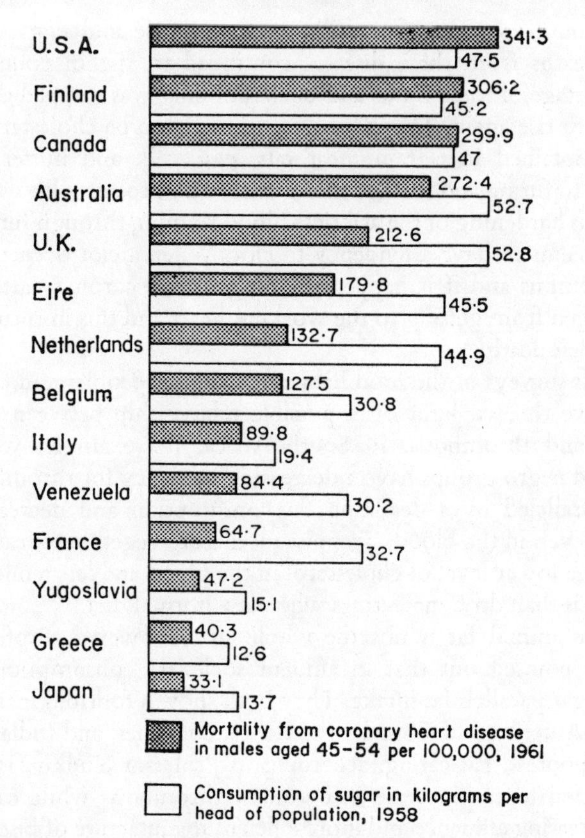

FIG. 20 Trends of mortality from coronary heart disease and sugar consumption in different countries.

After *Sunday Times*, July 1964.

TABLE 13

Influence of overweight on mortality in persons aged 45 to 50 years

Pounds overweight	*Increase in death rate over average* (%)
10	8
20	18
30	28
40	45
50	56
60	67

From Burn, H., *Drugs, Medicines and Man*, Allen & Unwin, 1962. (The figures do not necessarily mean that the deaths were caused by thrombosis alone.)

suspect in coronary disease: high blood pressure, diabetes, cigarette smoking, overweight (see Table 13) nervous stress and genetic factors. We cannot help our hereditary endowment but the affluent society in which we live does not encourage the simple, moderate living which may be the golden rule in preventing or at least cutting down this disease that kills over half the people in industrial countries, many of them in the prime of life, whose experience and skill we badly need.

THE GEOGRAPHY OF LUNG CANCER

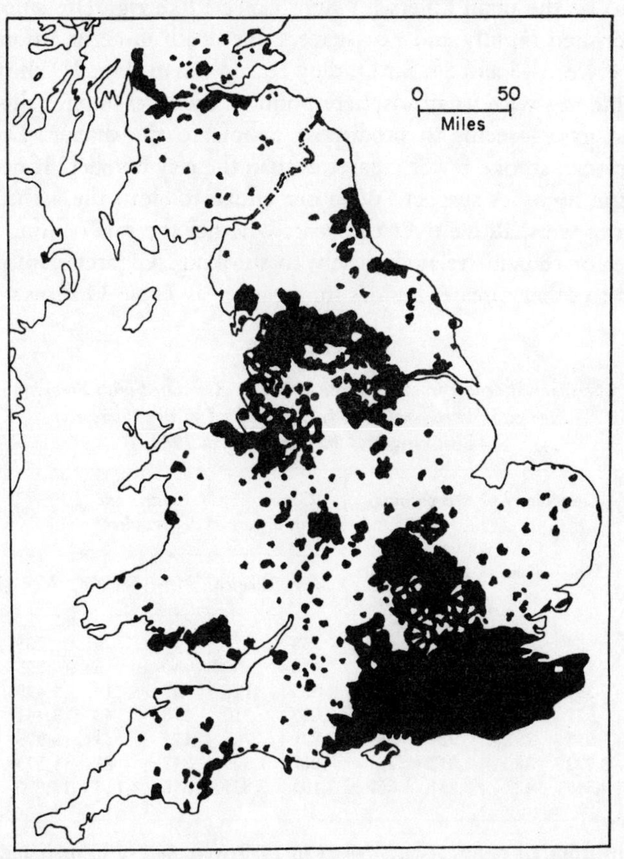

FIG. 21 Lung cancer (men) in Britain 1954–58, generalized. All parts in black have an incidence of deaths from this disease greater than the national average. This seems to be essentially an urban disease.

After Stamp, L., Dudley, *The Geography of Life and Death* (Fontana Library 1964).

SMOKING AND LUNG CANCER

Cigarette smoking is a common habit which began to be popular in 1871 when Virginia tobacco was imported, but it took over 30 years for it to become widespread. Pipe smoking, judging from the large numbers of seventeenth and early eighteenth century clay pipe fragments found in gardens, was a common habit from the middle of the seventeenth century but lung cancer was probably quite rare. Cigarettes seem to be the main killers for lung cancer; like cigarette smoking it has increased rapidly and now causes one death in eight among men aged between 45 and 54. Sir Dudley Stamp's map (Fig. 21) shows that town life too with its atmosphere polluted by soot, sulphur dioxide—exhaust gases—seems to predispose people to the disease. Does the countryman smoke fewer cigarettes than the city dweller? If not, then pollution becomes suspect. With our efforts to clean the air in town, and cities we shall see over the years whether the rate of lung cancer declines or remains related closely to smoking. Cigarette smoking is related to other diseases besides lung cancer as Table 14 shows:

TABLE 14

Number of deaths from Common Diseases Associated with Smoking Men & Women aged 30–64, England and Wales 1959, from Smoking and Health. *Pitman 1962.*

Age	*Lung cancer*		*Bronchitis*		*Coronary heart disease*		*Gastric and duo. ulcer*		*Total*	
	Men	*Women*	*Men*	*Women*	*Men*	*Women*	*Men*	*Women*	*Men*	*Women*
30–34	54	25	26	17	138	25	17	6	235	73
35–39	170	56	49	33	476	50	40	6	735	145
40–44	391	90	116	62	853	126	69	19	1,429	297
45–49	927	186	379	113	2,017	310	118	42	3,441	651
50–54	1,958	293	924	189	3,921	706	173	53	6,976	1,241
55–59	3,232	380	1,912	357	6,087	1,509	342	81	11,573	2,327
60–64	3,549	427	2,751	657	7,194	2,831	396	113	13,890	4,028

Deaths from all causes at ages 30–64 in 1959 were 84,296 in men and 51,989 in women.

The habit certainly shortens life. Only 15% of non-smokers aged 35 will die before the age of 65 as compared with 33% of 35 year olds who smoke 25 cigarettes a day.

DIABETES

One person in ten in affluent western countries may suffer from sugar diabetes (judging from a survey in Bedford), although most of these do not know they have it. Sugar diabetes is a disease in which the main blood sugar, glucose, does not enter the cells of the body properly. Consequently sugar piles up in the blood and leaks into the urine. Normally, when blood sugar levels rise, the hormone insulin acts to allow cells to absorb the extra sugar and so restores the sugar level in the blood to normal values. In diabetes, insulin for some reason does not act. Known diabetics can live normal lives by properly controlled diet and exercise, anti-diabetic pills and insulin injections, but many people are going about quite unaware that they have mild diabetes. There are perhaps 300,000 such people in Britain who, if they remain undetected and untreated, tend, for some reason, to develop hardening of the arteries and are liable to thrombosis. There is no doubt about the increase in known diabetics in affluent countries; in Denmark in 1927 about one in 1,000 was a diabetic but in 1946 one in 250 had the disease.

What are the reasons for this increase? Experts say that lack of exercise, increased sugar in the diet, being too fat and the increasing life span, are all factors tending to increase the numbers of people with this disease in the population. No less important is the fact that with the help of insulin, diabetics nowadays are able to live to reproduce and thus to propagate the trait.

THE CHANGING PATTERN OF DISEASE

Coronary thrombosis, stroke and lung cancer are three major killers of men in middle age or younger. Diabetes, if left untreated, can lead to death by the first two. In 1900 the major cause of death was still heart disease but tuberculosis and pneumonia accounted for 10% and 8% of deaths respectively. Now they account for less than 1% and 5% respectively.

An enormous change for the better has been in the reduction in infant deaths which made up 25% of all deaths in 1900. Infant mortality has now been reduced to about 2% of all deaths, thanks to the control and treatment of measles, whooping cough, diphtheria, scarlet and typhoid fevers and gastro-enteritis, all of which were serious killers in

1900. One example, diphtheria, will illustrate how immunization has cut down dramatically the death rate from this disease. Fig. 22 shows the trend of deaths from diphtheria in New York City and in England and Wales since 1900. It shows how great the fall in the death rate has been. First there was a gradual fall resulting from better living standards and medical care. Then in New York, about 1929–30, and in England and Wales about 1941–2, the graph pitches down and this is due to immunization of children against diphtheria. Another graph (Fig. 23) showing the incidence of polio in London illustrates how vaccine reduced the disease to insignificance.

The new killer epidemics of youngish and middle-aged men can be controlled but a great deal of the responsibility for their control lies with the individual and with the education of the individual. He can help to prevent his lung cancer by stopping smoking; he can ward off

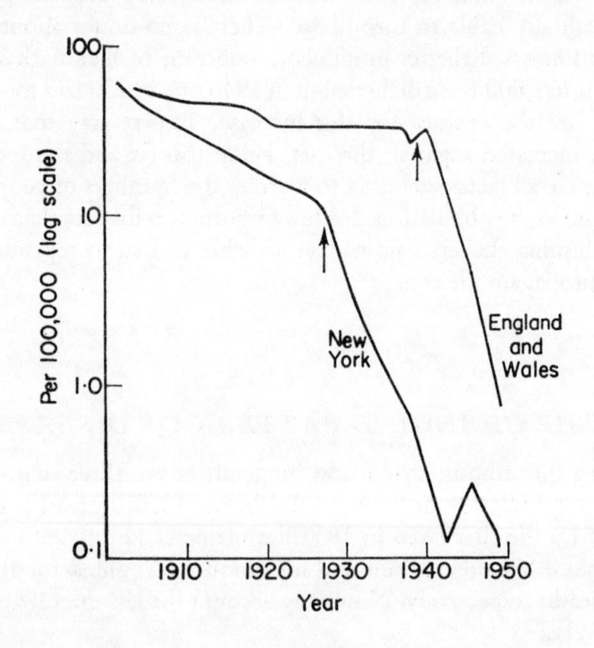

FIG. 22 The death rate per 100,000 from diphtheria in New York and England and Wales over the last 50 years. The arrows mark the year when immunization of 50% or more of children against diphtheria was first attained.

After Burnet, M., *Natural History of Infectious Disease* (Cambridge University Press, 1953).

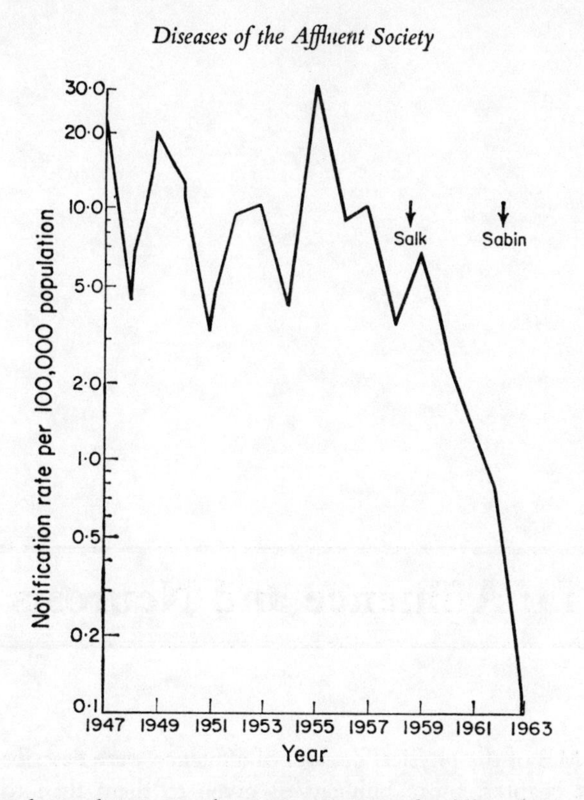

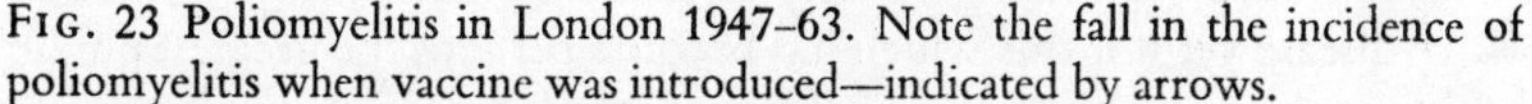

FIG. 23 Poliomyelitis in London 1947–63. Note the fall in the incidence of poliomyelitis when vaccine was introduced—indicated by arrows.

After *Report of the County Medical Officer of Health and Principal School Medical Officer for the Year 1963* (L.C.C., 1964).

his coronary thrombosis by using his legs rather than his car and perhaps by improving his dietary habits. Man was not made for a sedentary, affluent life but he can enjoy the amenities given by a good standard of living provided he realises that overfeeding, obesity and sedentary habits will, in the main, lessen his chances of enjoying a normal life span.

14: Affluence and Neurosis

SOME of the physical diseases of affluence were described in the last chapter; more publicity is given to them than to mental disease yet it is this which has become perhaps the greatest public health problem of our time. Half the hospital beds in this country are filled by the mentally ill. In fact, about 160,000 people are sick in hospital in this way but many, many more (and this is the hidden part of the iceberg) struggle along with their lack-lustre lives and neuroses. Indeed it is doubtful whether many families in Great Britain are free from neurotic illness of some kind, an illness which can cause much personal suffering and bring tragedy to family lives. A small survey by Lord Taylor and Dr Sidney Chave, carried out in a newly created town, 'Newtown', might well reflect in miniature the condition of mental health in this country and indeed in other affluent societies. Some figures from this survey are:

Of every 1,000 adults in 'Newtown' in 1959—
about 670 did not suffer any mental ill health
about 250 suffered from sub-clinical neurosis—'nerves', depression, undue irritability and sleeplessness;

about 80 suffered from 'neurosis proper' (i.e. anxiety states) arbitrarily defined as neurotic conditions which make people consult a doctor, of which,
about 6 were sent to see a psychiatrist, of these
about 2 were sent to hospital, of whom
less than 1 was psychotic, that is, suffering from a serious disease (as opposed to the minor mental illnesses above) which made them behave very abnormally. They are, frankly, mad.

These figures may not be representative of the country as a whole, but, if they are, then about one person in three has some neurotic symptoms. A large proportion of these suffer from mild neurotic symptoms, but about 1 in 12 (80 out of the 1,000 above) feels ill enough to take himself to the doctor for treatment. It is a difficult matter to know whether a neurosis is a product of the stresses of the environment or a part of the patient's constitution or the result of both. Certainly the hard cases of neurosis which need constant support throughout life form about 1 in 20 of the population and these are probably constitutionally neurotic. But there are many who belong to the sub-clinical group who are pushed into neurosis proper by environmental stresses such as money worries over hire purchase or keeping up with the Joneses or various other family strains. Taylor and Chave call these 'mixed neuroses' since they are due probably to inborn factors and environmental stress. Environmental neuroses are brought on in quite normal people by powerful environmental stress and generally clear up quickly once the stress is removed, if a worrying job is changed, for example.

What evidence is there that neurosis is a mark of an affluent society? It has been claimed that whereas in Great Britain 1 in 12 is a neurotic, in West Africa the figure is about 1 in 500. If the suicide rate is a symptom of the mental health of a country, Ireland, Italy, Spain, Egypt, and much of Africa should have better mental health than more affluent societies. For every 100,000 of the population, Ireland has 3·2 suicides and Egypt 0·1 against 11·3 in England and Wales and America. Even though the figures need interpreting with caution because suicide may be reluctantly recorded in Catholic and Moslem countries national variations are striking. In Great Britain alone there are probably about 50,000 attempted suicides a year and of these in 1963 about 5,721 were successful. Surveys, in fact, have shown that in cities attempts are six to ten times as frequent as successful suicides.

Why is it that there is considerably more neurosis and suicide in Britain and America than there is in Africa? It is not possible to give a cut and dried answer but perhaps African culture which is rich in tradition, rituals, and ceremonies gives the personality more room to move than do the hidebound conventions of western civilizations. Although the rituals of the Church persist here and in other rich countries, true religious conviction has declined in the West. Leisure time is often loafed away and there is very little desire to fill it with creative hobbies. In Britain the close-knit social networks of village and street life have tended to be destroyed with the advent of bleak new housing estates where people are often, at first, stand-offish and cool to neighbours because they are afraid of being gossiped about. Loneliness, social uncertainty and boredom, resulting from the upheaval from known ways of life often cause neurosis among wives. And, adding to the strain, is the cost of new furniture, the cost of travel to work or to see mother. Such cases parallel the neurosis in Zulu women which followed the destruction of the tribal system. In some of the developing African towns magical cults and healing shrines have multiplied in the last few years—'here in Britain we are more sophisticated; we take tranquilizers'. In short, neurosis and suicide seem to go with social disorganization: the break up of old behaviour patterns; the lack of community and family continuity and stability which leads to isolation and loneliness; the lack of standards and a philosophy of life.*

Contrary to expectation neurosis does not seem to have *increased* in Britain. Taylor and Chave's survey in Newtown showed that about one person in three suffered from mild neurosis. George Cheyne, an English physician in the eighteenth century, claimed that a third of his patients were neurotic. It looks as if neurosis, therefore, is a fairly constant feature of our society. What is certain is that there is nowadays more talk about mental illness and greater confidence in seeking advice.

Most surprising in the Newtown survey was that the rates for *major* mental illness (schizophrenia and manic-depressive illness) were below the national average whereas the rates for a decaying London borough were above the national average. This may be because the new town was a good society with its families healthy and happy and living in decent homes and its children going to schools in good buildings. Work was varied and close at hand and conditions of work good. The 'ordinary strains of industrial urban life were reduced to a degree not achieved in unplanned communities'. The implication is therefore that real madness can be prevented or made less severe by changing the social

* 'Suicide. The Terrible Escape', *Weekend Telegraph*, 14 October, 1966.

milieu and is not the result of inborn defects alone but of bad social conditions, which can make man alone and vulnerable. Even so the idea that good social planning reduces the incidence of neurosis is by no means proved.

DRUGS AND MENTAL ILLNESS

Millions of sleeping pills and tranquilizers are prescribed each year. Many are no doubt necessary but perhaps more outdoor exercise and creative hobbies would remove the need for them in many people. Some drugs certainly do help to 'raze out the written troubles of the brain'. There are drugs that lift depression and many that control anxiety or induce sleep. Others can control schizophrenia, a common and severe form of mental disease, which keeps over 40,000 people in mental hospitals, (one-third of the mental health in-patient population) at a cost of over £2,500,000 a year. It is not known how most of these drugs work (see Chapter 29) and perhaps it does not matter as long as they help to cure. But drugs with unknown side-effects need to be used sparingly especially those which relieve emotional distress caused by the changes and chances of life. Here they offer no lasting solution. Once the boundaries of such suffering are discovered and *explained* it may become bearable and curable. Nevertheless tranquilizing drugs have helped thousands of people back to normal lives. Fig. 24 shows how the effect of one of them, chlorpromazine, was to push down the numbers of mentally ill occupying hospital beds in England and Wales. Research over the next ten years will probably lead to the discovery of the causes of some severe mental illnesses like schizophrenia and manic depressive insanity. Already there is evidence that something goes wrong with the chemistry of the brain which drugs help to cure. The exploration of the interior of the brain, as J. B. S. Haldane has said 'will be as dangerous as that of the Antarctic continent or the depths of the ocean and far more rewarding'. Research on brain function and the relation between it and mental activity will probably uncover the effect of drugs, diet and hormones on the brain and its product—the mind.

Besides such 'physical research' it is necessary to find out a great deal more about the influence of the attitude of parents and the way in which they bring up their children on the development of the personality. The social relations of a patient and his culture, including the

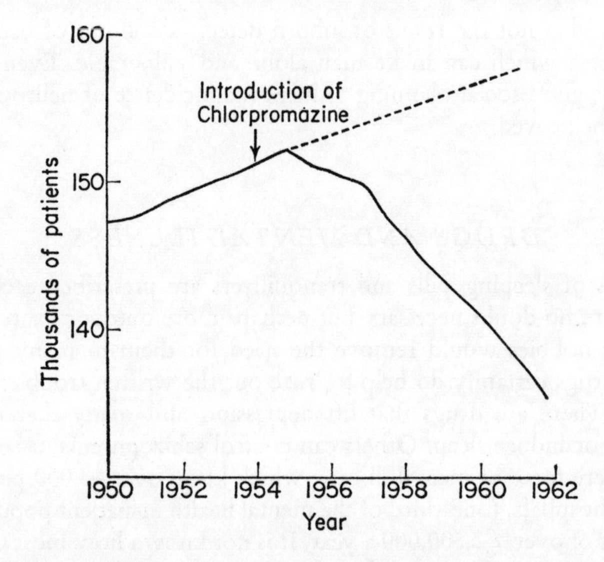

FIG. 24 The average daily bed occupancy (in thousands of patients) for mental illness in hospitals in England and Wales during the years 1950–62. The dotted line represents the possible increase of mental patients occupying hospital beds if chlorpromazine had not been discovered.
After Roy, J. R., *Science Survey B*, (Penguin Books, 1965).

things in which he believes, deserve close study as well. With this in view, psychiatrists who work almost entirely in hospitals or clinics need to pay more attention to the whole man as he functions in his environment. With the lengthening of life too it will be important to discover the physical, social and mental factors that cause mental illness in the old. Some of these are discussed in the chapter on old age.

15: The Iceberg of Disease

THE results of a survey in Bedford, reported in a previous chapter, showed that it is likely that one person in ten in this country may be an untreated diabetic. Other studies have shown that there may be 600,000 people with untreated bronchitis, 2,000,000 people over 45 with untreated high blood pressure, 250,000 with incipient glaucoma (a disease of the eyes that may cause blindness) 1,500,000 with untreated mental trouble and 400,000 women with untreated urinary infections that could lead to kidney disease, and 2,300,000 cases of rheumatoid arthritis. Rough estimates have indicated that in Great Britain there are probably more than 5,000,000 people with early unrecognized disease of one kind or another which might well be prevented if it could be detected soon enough. But how to detect and control these submerged, unobvious illnesses is a mammoth job that might well call for a revolution in the Health Service. The exposed tip of the iceberg, the recognized illness that brings a man to his doctor, is relatively small compared with the hidden mass of undetected disease, or 'at risk' people, in the population.

ROUTINE MEDICAL EXAMINATIONS

It is doubtful whether there are enough doctors to give all people over 40 a regular check-up in the hope of detecting hidden disease. It would perhaps lead to much wasted effort and time which the small force of doctors can ill afford. Probably the greatest value of a check up lies in its attempt to find out whether a person has shifted in his physical or mental health from his own individual pattern which has been obtained from past examinations. If a change is observed treatment can follow quickly.

The principle of the routine check-up has been applied to Child Welfare Clinics, the School Health Service, Ante-Natal Clinics (where the detection of an upward shift of blood pressure can send a woman to hospital for proper care and rest) and certain supposedly 'at risk' groups such as top business executives. Although the value of such a check-up for adults has been questioned on the grounds that it might make people worry too much about their health it is often of value in detecting the early symptoms of some diseases. Thus one study of 1,500 executives showed that 40% had some defect not previously spotted of which 90% could be cured. School medical examinations in 1958 brought to light 1,000 abnormalities among 15,000 children examined. Of these, 333 were of the ears, 188 of the heart, 244 were bone deformities, 225 chest defects and 64 epileptics. Ten percent of these had not been discovered at previous examinations. This evidence certainly points to the value of routine medical examinations. It is true of course that they might lead to a false sense of security. Provided, however, that a simple knowledge of the working of the body is part of the education of all and so long as the value and limitations of the tests are explained to the patient they could do nothing but good.

MASS SCREENING

Disease in the great mass of people, the submerged part of the iceberg, cannot be charted and controlled by conventional methods of diagnosis like the routine check-up. But certain leads towards mass screening have been given which are full of promise. Mass X-ray units are an excellent and cheap way of showing up early T.B., lung cancer and defects of the heart, which might not show in ill health for months or years, perhaps when it is too late.

The Bedford Mass Survey to discover early cases of diabetes was a model of team work by research workers, doctors, nurses, technicians and voluntary workers. It took five weeks and screened over 25,000 people. Every adult on the electoral roll, was circulated with a plastic pot for the urine sample and a letter stating the aims of the survey. The urine sample was then collected and tested for sugar (an indicator of diabetes) by technicians. Those suspected of diabetes were invited for a further blood sugar test. If this test was positive the patient was weighed and measured, the proportion of fat to bone estimated, blood pressure measured, and a medical history taken. From all these data it might be possible in time to pick out of the population people who are prone to the disease. It might also be possible to discover why they develop it. The striking fact emerging from the Bedford survey is that about one in ten people needed treatment for diabetes and it is likely that other towns and cities would give similar results.

Screening women over 30 to detect cancer of the cervix has shown that about three in 1,000 need treatment for cancer or early cancer. Cervical cancer kills 2,500 youngish mothers each year. It attacks 5,000 women in England and Wales each year and a further 8,000 show changes in the cells of the cervix which will eventually become cancerous. It is the commonest cancer in women. Luckily a method has been developed which shows up the cancer at an early stage. First some cells are taken from the cervix by a doctor and smeared on to a glass slide. The smear is stained by special dyes and examined under a microscope where abnormal and normal cells can be detected. Early cancer can be cured and the woman can still have babies. Many women are, however, too shy to ask doctors to carry out the necessary tests so a technique has been designed by which a woman can take a specimen of cells herself. The new aid is a plastic syringe like a fountain pen and with this she will be able to take a sample of cells and then post it to a diagnostic centre.

From these few examples it is clear that clinics need setting up in towns and cities for early detection of disease, such as lung cancer, diabetes, cervical and breast cancer, coronary artery disease and glaucoma. Laboratories are needed with proper apparatus and technical help, and time is required in which to carry out the tests and explain them. Most of the routine work (blood counts, urine analysis) need not be carried out by doctors but by trained technicians who would sort out the normal from the abnormal. Doctors could then devote their time to the abnormal findings. Indeed by discovering something about the habits and environment of say cancer sufferers some new ideas on preventive medicine may develop. Cancer of the cervix, for example, varies greatly from one social group to another. At the top of the social scale the risk is about half as much as for those at the bottom. Cleanliness has much to do with developing this form of cancer and this seems to be supported by a survey in New York which showed that cancer of the cervix was twenty times higher among Puerto Rican immigrants than amongst Jewish women and the difference was put down to sexual hygiene practised by Jews.

If 'preventive' medicine is not practised a patient who is diagnosed as having cancer of the cervix is perhaps beyond a cure. When coronary disease strikes, the blood supply to the heart might have gone beyond the point when a normal life can be resumed.

Preventive medicine need not be impersonal, for the family doctor is best placed to interpret the findings and give advice to the patient whose background he knows. Moreover he is in a position to influence personal and family behaviour into better habits of life.

SPECIAL GROUPS

Mass screening and standard medical examinations for special groups of people might well be very profitable in detecting disease. The chapters on heredity have shown that people with certain blood groups are prone to certain diseases. Broncho pneumonia in infants, duodenal and gastric ulcers, stomach cancer, pernicious anaemia and sugar diabetes have been found to be associated with certain blood groups and the possessors of such groups might be checked periodically for incipient disease. J. B. S. Haldane showed in 1938 that the chances of cancer attacking a person are much greater if one of his parents suffered from the disease than if neither parent was a sufferer. Out of 953 sons whose fathers died of cancer, 14 developed the disease. In a random sample of 95,300 men and boys of similar ages only 85 on average developed cancer instead of 1,400 expected if the frequency were similar to that of the first group.

Such 'at risk' groups need screening by routine examination to identify early stages of cancer and other diseases.

Nationalities show great differences too in their proneness to cancer. Cancer of the gullet is much higher in all foreign-born white men than in American-born whites and Polish, Czech and Irish immigrants show the highest rate. Habits of life—foods, alcohol consumption and certain types of cooking as well as smoking may be the causes.

BODY BUILD, DISEASE AND PERSONALITY

'Fat and merry, lean and sad' is a commonly held notion about the link between body build and personality. Indeed such views are very old; Hippocrates, the father of medicine, thought that the bean-pole (linear) type of person was more prone to tuberculosis while the broad type (lateral) tended to suffer from heart attacks and stroke. As in so many old sayings there is perhaps a grain of truth in this. There is some evidence that tall people with a low weight and with long narrow chests, provide a fertile soil for the T.B. germ; heart attacks and strokes happen most often in people of high weight for their height. They occur more often too, in muscular he-man types, especially those who do not take enough exercise. Apparently they need this exercise more

TABLE 15

Body Build and Disease. After Linford Rees, 'Physique, personality and disease' in Science Journal, *June 1965.*

Physical and mental traits	*Linear build*	*Lateral build*
Personality characteristics	Introvert Inhibition, anxiety and depressive tendencies More persevering Quicker tempo	Extrovert Sociable and cheerful Less persevering Slower tempo
Neurosis	Anxiety and depression Schizophrenia	Hysteria Manic depressive illness
Physical disease	Duodenal ulcer Pulmonary tuberculosis Early diabetes	Gall bladder disease in women Coronary disease Arterial disease Late diabetes

than do less muscular but equally lazy people. Diabetes is also said to be related to body build. People who suffer from a form of the disease that appears early in life are usually linear types while those who develop it later are usually lateral types. And everyone recognizes the duodenal ulcer type who is invariably linear.

As for mental disease, linear people are more likely to suffer from schizophrenia than those with a lateral type of build who are more prone to manic-depressive illness. Fat (lateral) children are more outgoing and confident than their linear companions who are often shy, anxious and sensitive and do not easily communicate their feelings to other people, while muscular children are generally good mixers. Table 15 summarizes some of the personality traits and diseases which *tend* to develop in linear and lateral people.

It is very likely that the genes responsible for the body build of a person also influence personality and mental and physical illnesses. A great deal more research is needed in this field, however, before proneness to a disease can be predicted accurately (if indeed this ever can be done) by examining physique alone. Like computer medicine and blood groups, body build tells a doctor something more about his patient. A doctor will always need to identify symptoms, decide which are important and give the treatment indicated. Body build might eventually be recorded on an individual's health card so that diseases that are likely to develop can be watched for and checked.

16: Diseases of Poor Countries

UNDERDEVELOPED areas like Africa and India have patterns of disease which differ from those of the rich western countries. Their illnesses are perhaps similar to those experienced by British people at the time of the Industrial Revolution (e.g. typhoid and tuberculosis). For these countries medical textbooks are probably too up to date and quite inappropriate, for they deal with the problems of health and disease which prevail in affluent countries. The difference between the disease patterns of richer countries such as U.S.A., Canada, and England, and the poorer ones, is probably related to two other important differences—the standard of nutrition and the age structure of the populations. Here and in America and Canada there is a large proportion of adult and old people and a high rate of disease due to old age, overeating and lack of exercise. In poor countries food is in short supply and good nutrition, particularly protein nutrition, is a keystone for survival. Fig. 25 showing the daily calorie content of food eaten in different parts of the world and Fig. 26 showing the daily intake of protein in grams, illustrate dramatically the truth of the saying that 'half the world goes to bed hungry'.

In India, although the birth rates are very high, about half the

WORLD INTAKE OF CALORIES 1957–61

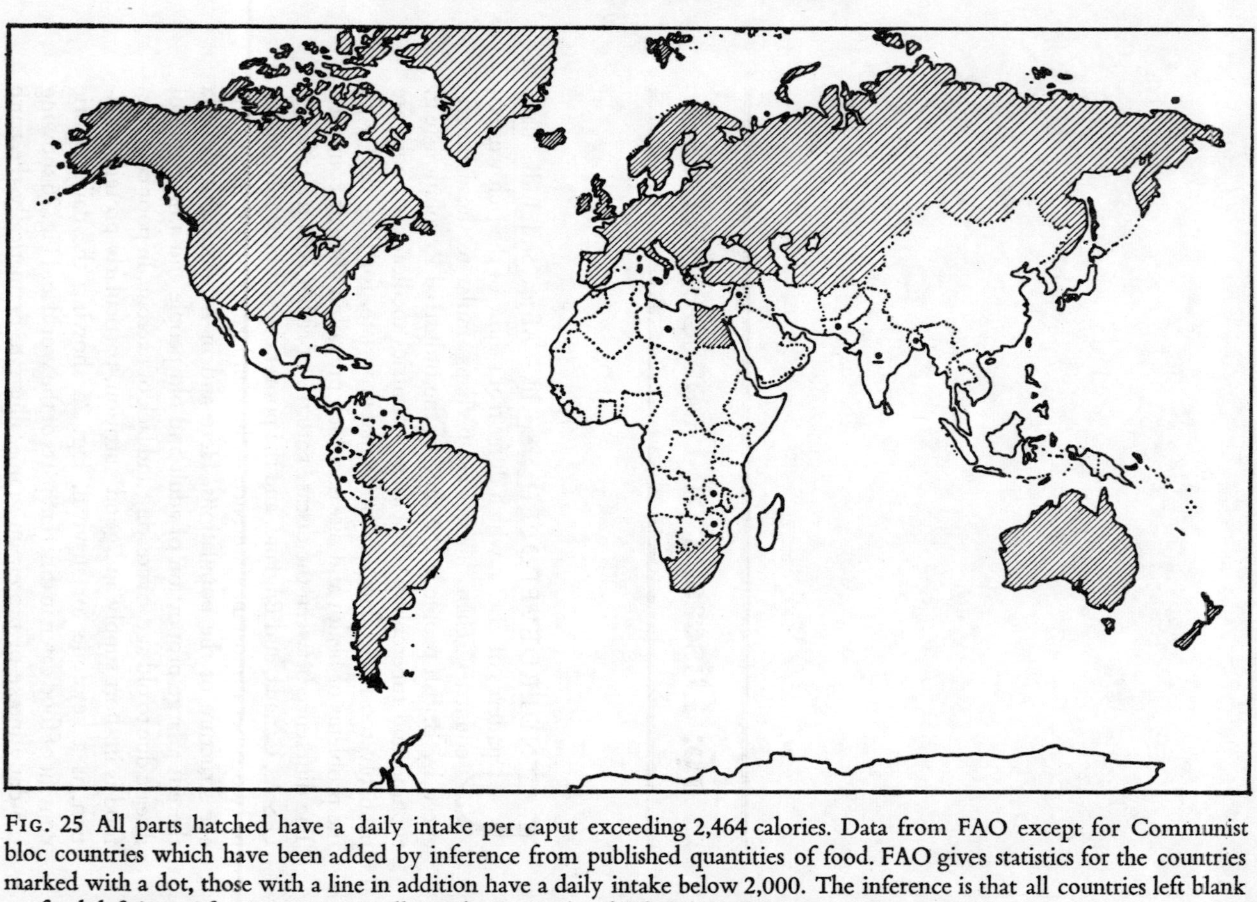

FIG. 25 All parts hatched have a daily intake per caput exceeding 2,464 calories. Data from FAO except for Communist bloc countries which have been added by inference from published quantities of food. FAO gives statistics for the countries marked with a dot, those with a line in addition have a daily intake below 2,000. The inference is that all countries left blank are food deficient. After Stamp, L., Dudley, *The Geography of Life and Death* (Fontana Library, 1964).

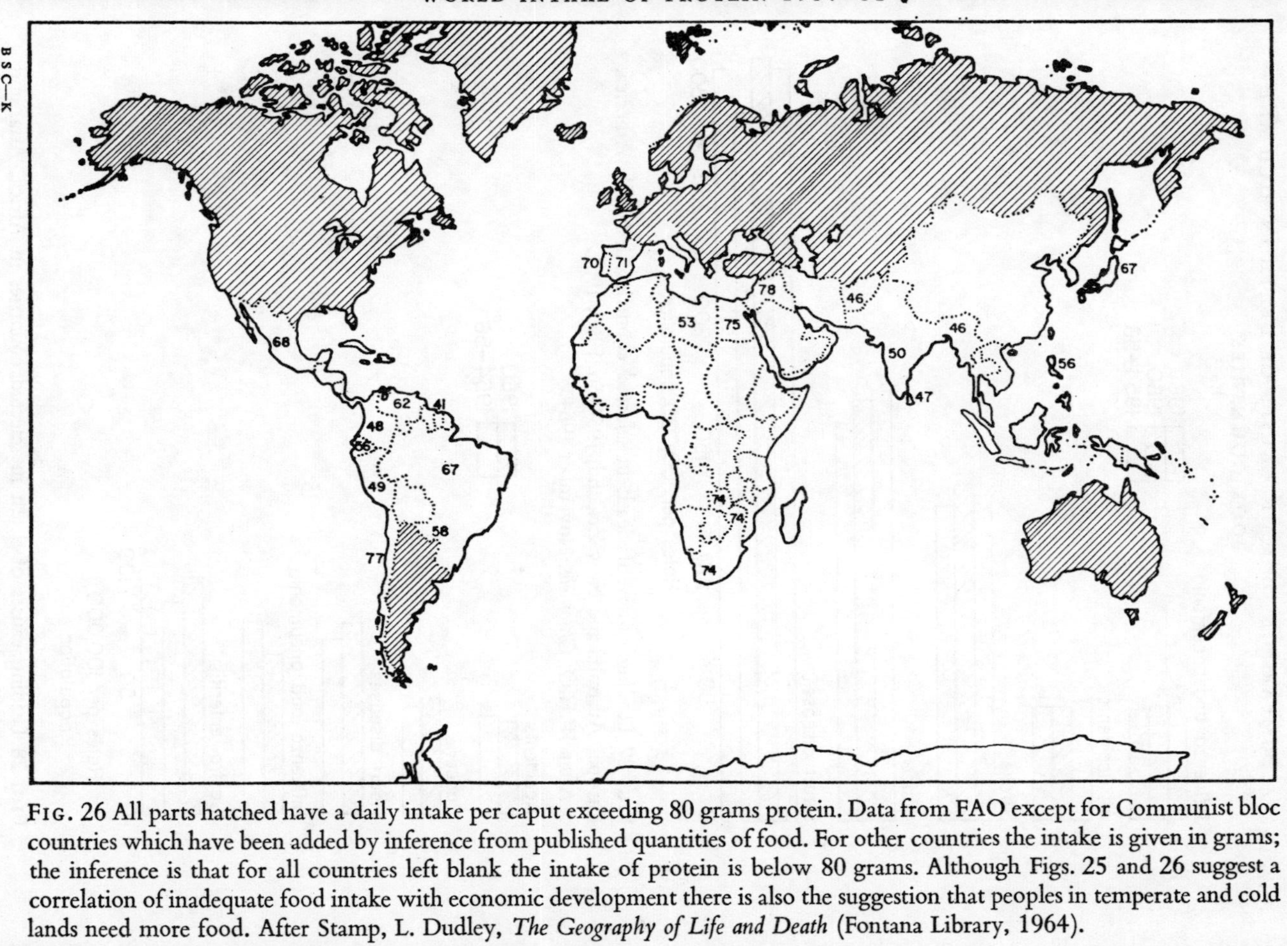

FIG. 26 All parts hatched have a daily intake per caput exceeding 80 grams protein. Data from FAO except for Communist bloc countries which have been added by inference from published quantities of food. For other countries the intake is given in grams; the inference is that for all countries left blank the intake of protein is below 80 grams. Although Figs. 25 and 26 suggest a correlation of inadequate food intake with economic development there is also the suggestion that peoples in temperate and cold lands need more food. After Stamp, L. Dudley, *The Geography of Life and Death* (Fontana Library, 1964).

A COMPARISON OF MAIN CAUSES OF DEATH IN RICH AND POOR COUNTRIES

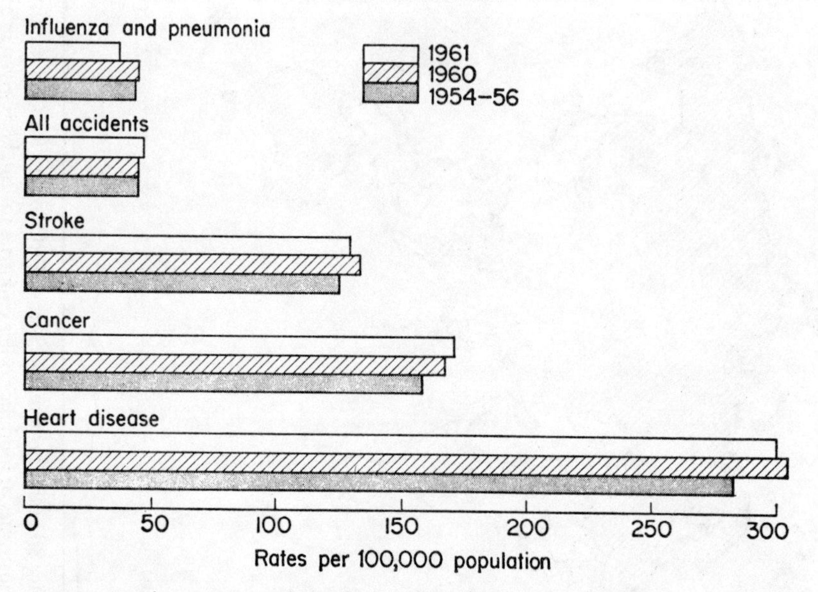

FIG. 27 Leading causes of death in selected countries in North America, Europe, Australia and New Zealand, 1954–56, 1960, 1961.
After *W.H.O. Chronicle*, November 1964.

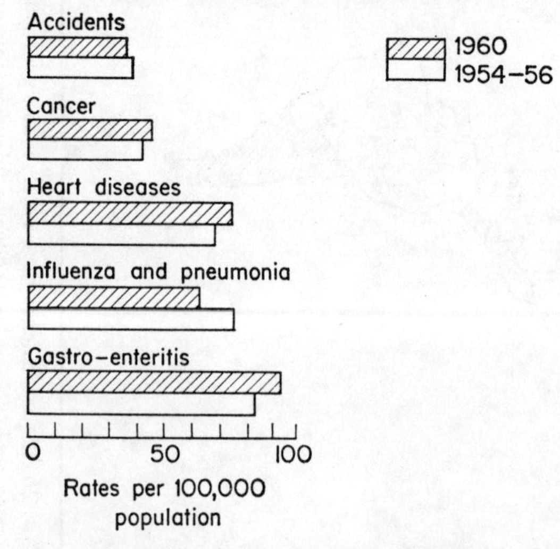

FIG. 28 Leading causes of death in selected countries in Africa, Asia, and Central and South America, 1954-56 and 1960.
After *W.H.O. Chronicle*, November 1964.

children do not survive beyond the age of five, while in Zulu communities in Africa about 35% of babies are born dead. Many of these deaths are caused through malnutrition of the mother while many of the deaths of young children are due to gastro-enteritis in the summer and measles and pneumonia in the winter—diseases that well-fed, robust children can resist more easily. Malaria, syphilis, tuberculosis, kwashiokor (a serious and often fatal expression of protein malnutrition) and epidemics of typhoid, cholera, and plague kill many older children. Figs 27 and 28 illustrate the differences between poor and rich countries in the main causes of death.

It is the young adult who predominates in Africa, India, and Middle and South America (see Fig. 29); old people are rare. In India in 1950, for example, the average expectation of life was 32 years while in England and Wales it was 65. The sad truth is that most of these deaths are caused by illnesses which could be cured, but lack of money and two of its offshoots, lack of a balanced diet and lack of education, allow diseases to flourish that have been stamped out in the West. In parts of Africa, for example, prejudice dictates that pregnant women must neither drink milk nor eat eggs for fear of having a timid son. This denial of protein-containing food is damaging to the health of both mother and child.

Whether or not the modern western malady of stress is rare in countries like India and Africa is an open question. Violent changes in customary ways of life are common forerunners of stress, and such changes must certainly be taking place in Africa and other 'poor' countries at the moment. The history of the Zulu people is a case in point.

In the nineteenth century large areas of South Africa were occupied by the Zulus. The main domestic job of the men (warriors by instinct and upbringing) was tending herds of cattle while the women did the hard, menial work on the land. The women, in fact, were at the bottom of the peck order; at the top was the Zulu king. Now there is no longer a Zulu king; his authority has been taken over by the white man. Zulu lands have been reduced and, to make ends meet, most men find work in mines and factories for most of the year. In short, the family unit and the larger tribal unit have been broken up and old traditions and regulations are a thing of the past. Such a gross shaking up of old established patterns, leaving the individual in an unfamiliar wilderness provides a fertile source of stress which can be observed in the Zulu.

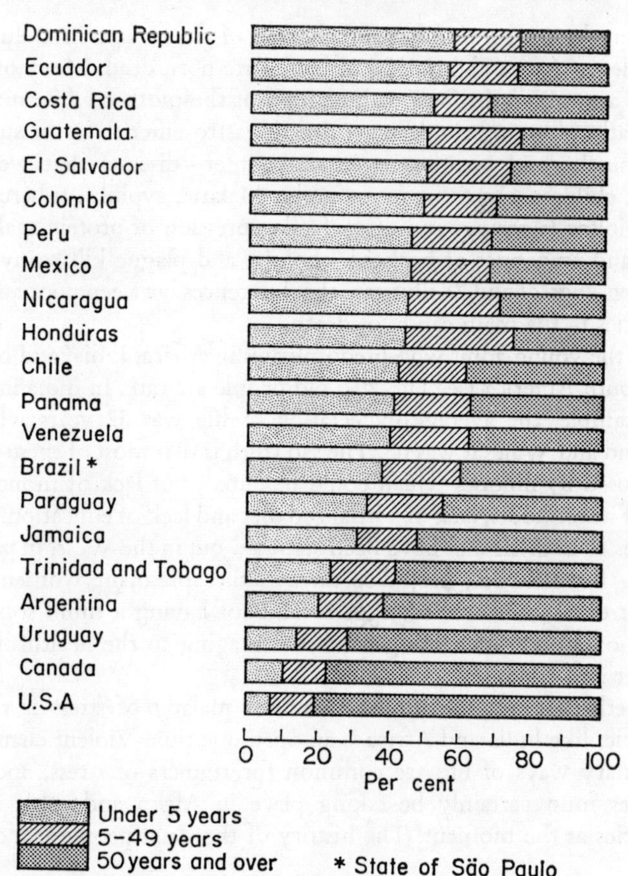

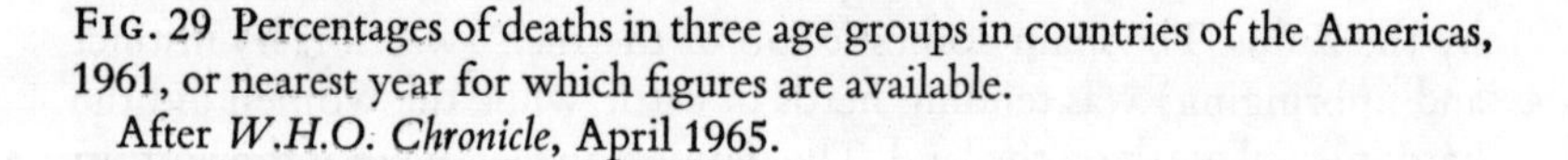

FIG. 29 Percentages of deaths in three age groups in countries of the Americas, 1961, or nearest year for which figures are available.

After *W.H.O. Chronicle*, April 1965.

It remains to be seen whether the new diseases of the West—thrombosis, lung cancer, diabetes—apparently not so frequent in Africa and Asia—will become more universal as western standards of living spread. The word 'apparently' is used advisedly, because although peptic ulcers appear to be rare among Africans, post-mortems reveal ulcer scars as frequently as in Britain; the Africans have carried on, because their economy has no place for the unproductive, bed-ridden patient.

The much grimmer spectre of over-population is just around the corner. As people are saved from needless death, as infections decline and general health is improved, more and more mouths are left to be fed. This subject of population forms a separate section of the book, and will not be considered here.

17: What People Eat

IN the last chapter it was stressed that good nutrition is the foundation of good health, What people eat and the way in which they get their food is a complex, fascinating and sparsely documented story. One fact which needs to be stressed at the start is that the human digestive system is very flexible and can cope with a wide variety of foods. This ability forms one of the most far-reaching adaptive properties of man. Table 16 illustrates this 'metabolic flexibility' of the body which is able to use the three energy-giving substances, fat, carbohydrate and protein in a great variety of combinations and from a wide variety of sources.

Despite the flexibility of his digestive system, man's palate is conservative. It is unlikely (but not disproved) that his palate is conditioned by genetic factors; rather it is governed by taboos, beliefs and what is available in the area.

An amusing and very local example of the conservatism of the British palate is shown by what a cross section of the British population thought to be 'the perfect meal' in 1947 and 1962 (see facing page).

Even after 15 years of good living the affluent British society still chose a meal almost identical with that chosen by the austere Britain of 1947.

TABLE 16

Average daily intake, per head, of fat, carbohydrate and protein, and their sources. From Human Biology, *Harrison et al., O.U.P., 1964.*

Food	*Sources*	*U.K.*		*Kikuyu*		*Eskimo*		*Barbados*	
		g	*cal*	*g*	*cal*	*g*	*cal*	*g*	*cal*
Fat (9 cal/g)*	Fats, oils, animal foods	110	990	22	198	162	1,458	63	567
Carbohydrate (4 cal/g)	Cereals, starchy roots, fruits, sugar	400	1,600	390	1,560	59	236	416	1,664
Protein (4 cal/g)	Meat, fish, milk, pulses, nuts	100	400	100	0	377	1,408	45	180
Total		—	2,990	—	2,158	—	3,102	—	2,411

* The energy value of food is measured in calories. A calorie is the amount of heat needed to raise one kilogram of water through 1°C. The number of calories needed from food to keep the body going depends on its size, the severity of work and the climate. The colder the climate the greater the number of calories needed.

'THE PERFECT MEAL'*

1947	*1962*
Sherry	Sherry
Tomato soup	Tomato soup
Sole	Sole
Roast chicken	Roast chicken
Roast potatoes, peas and sprouts	Roast potatoes, peas and sprouts
Trifle and cream	Fruit salad and cream
Wine	Wine
Coffee	Coffee
Cheese and biscuits	Cheese and biscuits

* From *Plenty and Want* (see Bibliography).

We are, as yet, very ignorant about what people eat and of the basic calorie needs, of selected groups of people, If these were known we should be able to estimate the 'calorie cost' of food-getting activities such as collecting firewood, pounding rice, rounding up cattle and so on, and thus be able to plan accordingly on how to feed and improve the living of primitive communities.

The food of the Eskimo is perhaps the extreme type of carnivorous diet. It is likely that it illustrates the kind of food that the hunters of Palaeolithic and Mesolithic times existed on. His food is nearly all protein and fat, mostly obtained from the sea. Sometimes 20 pounds of meat is eaten at a single meal, but meals are chancy and when fog comes down the Eskimo may face starvation. Cranberries and blueberries, roots and young leaves of the dwarf willow are included in the diet and give him the necessary vitamins. Eskimos are healthy, the protein-deficiency diseases of Africa, India and Latin America mentioned in the last chapter are unknown. By contrast the diet of a man from Southern India is vegetarian; rice forms just about 85% of the diet of the poor. Spices and chutney add savour and pumpkins, chillies and tomatoes add colour and some vitamins. Protein is low because the slaughter of animals is forbidden by Hindu custom; if it were not, the rich fisheries of the Indian Ocean could be drawn on, cows could be killed, and rats, which eat up grain stores could be exterminated. In this diet, then, we see religious taboos, poverty and ignorance acting together to produce one of the least nourishing diets in the world, which, as we have seen, shortens life and produces a high infant mortality. On the other hand the Sikhs in northern India live on a diet of meat, milk and milk derivatives and the toughness and superior physical resistance to disease of this people is directly related to their high protein diet. A graphic experiment by McCarrison with rats fed on the Madrassi diet and on the Sikh diet produced an average weight of 155 g for those fed on the former while those fed on the latter diet weighed an average of 255 g.

Much of the diet of Latin America and Africa is starch. In the former, cereals, beans and potatoes dominate but eggs and fish, although rare, give a little protein balance. Beef and poultry provide the meat source but are expensive; a skilled labourer in a city in Chile must work for about three hours a day to earn enough to buy a kilogram of beef. In America the same quantity of meat would be earned in 20 minutes as Fig. 30 shows. The situation is similar with other foods. It is not surprising that protein deficiency forms a serious problem facing maternal and child health services in a good many of the developing countries. In peasant communities in Gambia they live mainly on starch in the form of rice although millet and sorgum is also used. Eggs and milk are uncommon but small quantities of meat, dried fish and beans give some protein balance. Various fruits and leaves in season add vitamins, but the picture is one of protein hunger. Most primitive societies such as the Australian bushmen have an excellent knowledge of the country and food values. An unskilled observer might think that such diets were grossly deficient in protein until he realized that

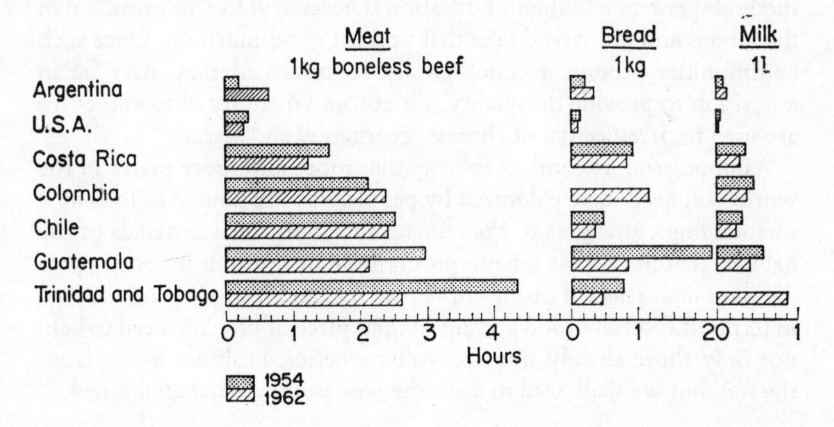

FIG. 30 Working time required to purchase three basic foods (1 kg. of meat, 1 kg. of bread, 1 litre of milk) in selected countries of the Americas, 1954 and 1962, based on average earnings of skilled labourers in selected cities.
After *W.H.O. Chronicle*, April 1965.

protein and fat in the form of grubs, caterpillars, frogs, snakes, lizards, locusts, ants and termites add to their value.

Sufficient has been said in this chapter and the last to show that much of the world is protein hungry while other richer parts of the world, such as urban America and Western Europe, eat a huntsman's diet of protein after a day at the desk. We must be cautious of generalization however. Even now in a rich country like Great Britain, rickets, a disease of the industrial revolution is apparently still with us: out of a sample of 200 children in Glasgow, 27 showed signs of rickets or old rickets, and out of a sample of 600 seventeen-year-old army recruits nearly half had poor physique or showed signs of old rickets. Most of them came from large Northern cities and their condition was probably due to malnutrition in childhood—cakes and sweets rather than vegetables and fruit and perhaps a biscuit for breakfast instead of protein.

In spite of protein shortage it is very doubtful indeed whether the juice of leaves pressed out by machines and made into protein-rich 'steaks' or 'cakes' would appeal to the human palate (see Chapter 20). It is far more likely, if observations on human nature are anything to go by, that people would like to eat more of the food they eat now. And this is likely to happen in the developing countries as science and its methods penetrate to them. Education is necessary, too, to break down the taboos and old wives' tales that prevent good nutrition. Once such communities become technologically sophisticated they may be in a position to provide the quality, variety, and quantity of food that we are used to, irrespective of climate, geography and season.

As population expands it follows that more and more places in the world will need to be colonized by people who are unused to their new surroundings, strangers to the climate and nutritional demands of the habitat. It follows that a basic programme of research is necessary to discover how selected groups of people live and how they are adapted in terms of food and food getting to their place of life. We need to help not only those already there to wrest a better, healthier living from the soil, but we shall need to assist the new people to get established.

18: Some Myths about Food and Health

IT is unlikely that the food of the Eskimo or the poor Indian will suffer from the addition of chemicals to make it more palatable. This however has been the fate of western diets for over 150 years. Indeed the food faddist in our society often declares that our food is so impure and 'chemicalized' that it is bad for our health and that a great many of us eat less well than did our great, great grandfathers a century ago. It is true, as we have seen in Chapter 13, that certain fats and sugars, along with other factors, may help to cause thrombosis while changes in diet almost certainly play some part in causing diabetes, peptic ulcers, appendicitis, and disagreeable conditions such as constipation, biliousness and indigestion. There is always the possibility too of being poisoned over long periods and even developing cancer from food additives such as colouring matter, preservatives, emulsifiers and anti-germ agents. In addition chemical residues (left in meat) of growth-promoting hormones and tranquilizers given to cattle and poultry to fatten them or to calm them when slaughtered, have given some cause for uneasiness among doctors and the general public. The cry of the food faddist acts as a useful corrective to over-enthusiastic food technologists. Indeed there is a need to adopt a very strict approach to those additives which serve no useful purpose except to make food more tasty or look more attractive.

The effect of food technology is to alter food from its original simplicity and this is absolutely necessary if a large urban population is to get a fair share of fresh food to eat at any season of the year in any place. If canning, freezing and other methods of preservation had not been developed we should have to live on stale and unattractive foods for most of the year with deleterious effects on health. But with sophisticated foods may come a multitude of dangers. They do not have the 'hard-bite' of natural foods; often, as in biscuits, they are concentrated, pappy and sugary. They play havoc with children's teeth because they encourage the growth of caries—producing bacteria in the mouth which lead to tooth decay. Four million teeth were extracted from English children in 1965 because of decay. One interesting study of the people of Tristan de Cunha showed that their teeth were in excellent condition before the arrival of a canning factory on the island when their teeth started to decay. The canning factory brought with it the usual canteen which provided sugar, white flour and sweets in large quantities and these the islanders took a liking to. Their diet thus changed from a primitive to a sophisticated one and brought with it the expected increase in tooth decay.

Obesity, a cause for shame in affluent societies, is encouraged by food sophistication. It also has a genetic component: Pickwick's fat boy could not help being fat. Obesity brings with it increased likelihood of premature death as we have seen in Chapter 13 and a train of ills, some of them lethal: high blood pressure, stroke, thrombosis and diabetes. Natural foods such as fruit and vegetables containing roughage in the form of cellulose make the stomach satisfied before too much has been eaten, but biscuits, sweets, cake and ice-cream are without this built-in restraint and large numbers of calories are eaten before hunger is satisfied. Added to this is the intense bombardment of our will to resist by the advertising machine of the food industry; glossy magazines and luxury displays of food in shops whip up our appetites above our needs.

All this does not appear to be a very healthy situation, but it needs to be said at once that the beneficial effects of twentieth century food sophistication on life expectation and the health of children far outweigh any adverse effects. There is always a risk in using food additives and their use needs to be, and is, watched vigilantly by public analysts. If the risk is so small that it is apparent only to the professional 'viewers with-alarm' then at least it should be balanced against the economic advantages of the additives in question. Similarly the dangers of chemical residues left in meat need to be weighed against the demand on farmers to produce cheap meat, milk and eggs.

Most of us would wish the food industry to produce foods nearer to the natural and simple but we need to be on our guard against the cries of food faddists who say that there is hardly an honest food left to buy and that all the goodness has been processed out of it. It is excellent that the food fad should act as a gadfly to the over-enthusiastic food chemist, but as Dr John Burnett has recently shown us* it is no overstatement to say that mass-produced food is purer today than it was a century ago. In the 1850's for example, bread provided, one writer said, rather a crutch to the grave than a staff of life for it was often whitened with alum so that the baker was able to pass off a loaf made with cheap flour as a top quality one collecting, of course, the higher price. Milk was watered shamelessly and this together with the poor bread must have lowered children's resistance to disease. As in the developing countries today, infants in early nineteenth-century England died like flies. Tea was faked on a large scale; it was made sometimes from the leaves of ash, sloe and elder which were curled and coloured on copper plates. Coffee was adulterated, often with ground-up baked horse livers and beer was watered then 'strengthened' by drugs (some poisonous), hartshorn shavings, orange powder, caraway seeds, ginger and capsicum, all cheap substitutes for real malt and hops. Cheese rind was sometimes coloured with vermilion and red lead. Sweets were a veritable concentration of poison; some were coloured with chromate of lead, others with red lead and some with arsenite of copper. The sugar base was sometimes adulterated with hydrated sulphate of lime and even white lead was used to put the colour on. It is no wonder that chronic gastritis was one of the commonest diseases of urban populations in early Victorian England and was no doubt due to slow poisoning by traces of lead, copper, mercury and arsenic piling up in the system.

Such mass adulteration was a natural outcome of the rapid growth of urban populations in the early nineteenth century, populations that were dependent on others for their food, and the native wit of man saw a way of making money. There are a few earlier records of food

* *Plenty and Want*, Nelson, 1966.

adulteration but they are insignificant compared with what happened in England a century ago. This widespread trickery and poisoning puts into perspective the recent isolated case of Blackpool rock coloured with Rhodamine B which in sufficient quantity might cause cancer. It was held to be no defence that an individual would have to eat a hundred tons of the rock for the concentration of the chemical to become critical. Some years ago the late Sir Edward Mellanby demonstrated that bread which had been 'improved' by agene caused hysteria in dogs; the bread did not seem to affect humans but agene was rightly banned. Such cases, although rare, demonstrate the need for eternal vigilance by public health authorities. Education, too, must play its part. Does the average housewife know, for instance, that 'orangeade' may be made without oranges and that shop meringues may be made out of cellulose material and not the expected egg white? This subtle form of 'legalized adulteration' needs watching by professionals and by a critical public.

THE ILL-FED LABOURER

We may eat purer food now (perhaps too pure—our bread is worse for this) but how much better fed are we as a nation than a century ago? Much better. A great deal of Britain then, as far as nutrition and health are concerned, was in the same state as India is now. The town and city dwellers were, we know, seriously undernourished, but until Dr Burnett told us it was not commonly known that the rural populations were often worse off than their brothers in the towns and the farm labourer, whom we might have thought kept his pigs and hens for bacon and eggs and got his milk from his employer, was probably the worst fed of all workers in the nineteenth century:

> He used to tramp off to his work while town folk were abed,
> With nothing in his belly but a slice or two of bread;
> He dined upon potatoes, and he never dreamed of meat
> Except a lump of bacon fat sometimes by way of treat . . .

Bread, then potatoes, with meat as a luxury. That was the lot of the labourer right up to 1914 when even then his diet was surprisingly low in meat, milk, eggs and butter. True there were the rich and the yeoman farmers and it is to them we hark back when we long for the 'plenty' of the Georgian or Victorian table with loins of beef and saddles of mutton washed down with pots of home-brewed ale. We tend to think, one might say, of Gainsborough rather than of Hogarth.

The gap between the rich and the poor was vast; it was just as great as it is today, not only in the developing countries but in Spain, Italy, and even America. Progress—agricultural, medical and social—has slowly closed the gap between what the worker eats and what the professional man eats today. Now it is not the social class that matters in standards of nutrition but the size of the family. Even in the highest social class (gross weekly income of head of household £34 or more) there is some protein deficiency in families with four or more children.

We have seen then that the food of the nineteenth century was far more impure than our own, and that trickery and fraud were widespread. We have seen also that the differences in quality of diet then were very great and that the worst fed people were surprisingly the ones who worked to produce food for others. In the middle of the twentieth century our children are taller and heavier than ever before and our life expectation is longer, benefits which have accrued as a result of good diet. Not only is there enough good food to eat but it is enjoyed by most of us, and not just by the privileged few as it was in Victorian times. We need to look now to the waste of food that goes on and to some of the dangers to health of over-feeding.

19: Horizon Medicine

It is funny, you will be dead someday.
E. E. Cummings: Sonnets.

THE previous chapters have been concerned with some diseases of rich and poor societies, some ways of preventing disease, and something of man's varied and changing diet. This chapter looks further at some human problems whose solutions may lie on or over the horizon. It seems that the answers to problems like quelling the population explosion, preventing cancer, and the development of spare-part surgery are quite near to us. The use of computers may soon be possible too, to store the diagnostic, but unfortunately mortal, skill of distinguished doctors for use on patients as yet unborn, in Cambridge or Calcutta. Other achievements like the extension of youth or predicting human ability and skills accurately, or breeding men for space travel, are miles away over the horizon.

THE EXTENSION OF YOUTH

Prolongation of an active and healthy life is probably one of the most ancient and elusive of human ambitions. Dr Alex Comfort, who has worked for many years on the problems of ageing, pointed out that

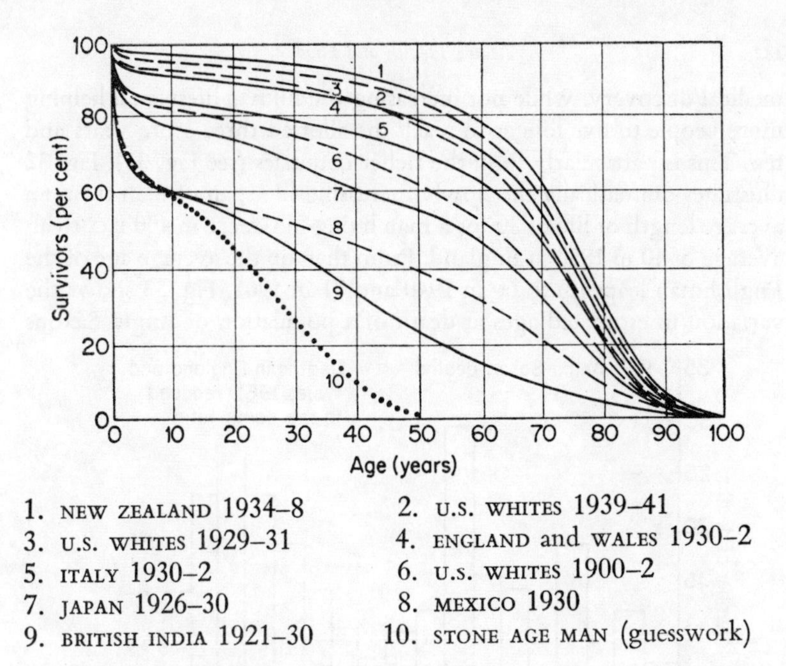

FIG. 31 Human survival curves, all for females, show the effect of improvements in living conditions.
After Comfort, A., *The Process of Ageing* (Weidenfeld & Nicolson, 1965).

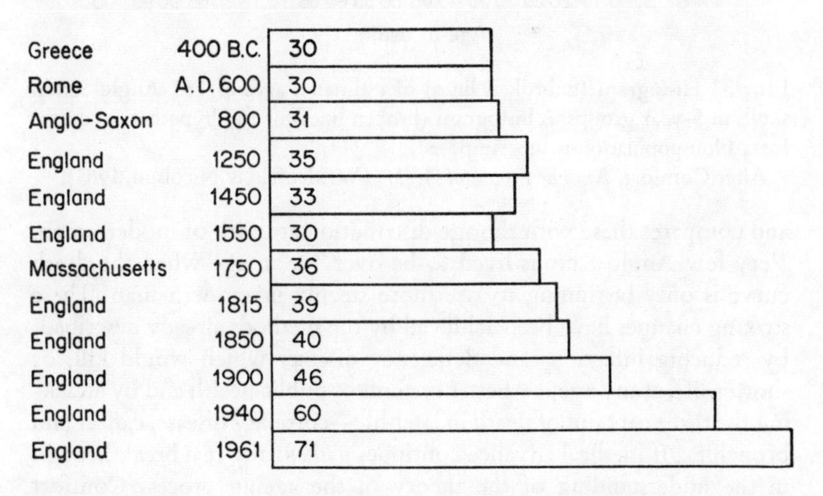

FIG. 32 Diagrammatic representation of the average length of life from ancient to recent times.
After Comfort, A., *The Process of Ageing* (Weidenfeld & Nicolson 1965).

medical discovery, while not increasing the human lifespan, is helping more people to live longer to reach the allotted three score years and ten. This is particularly so in the richer countries (see Fig. 31). Fig. 32 illustrates dramatically the slowly increasing life span of man from an average length of life of 30 for a man living in Greece in 400 B.C. to an average of 40 in 1850 in England. From then on the average age of the Englishman leaps up to 60 in 1940 and 71 in 1961. Fig. 33 shows the variation in estimated ages at death of a population of Anglo Saxons

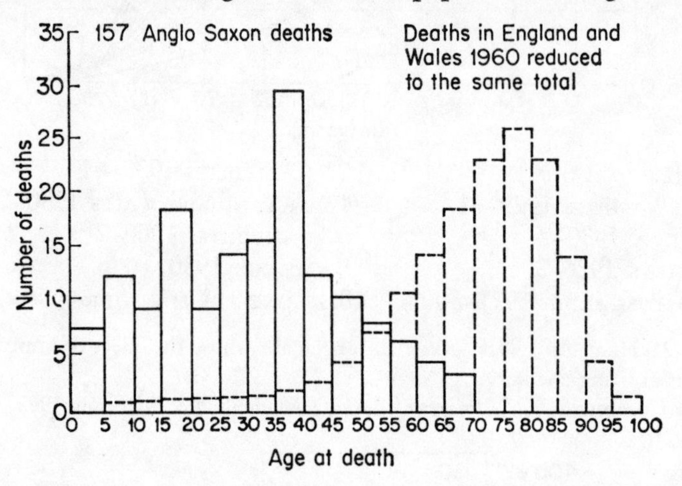

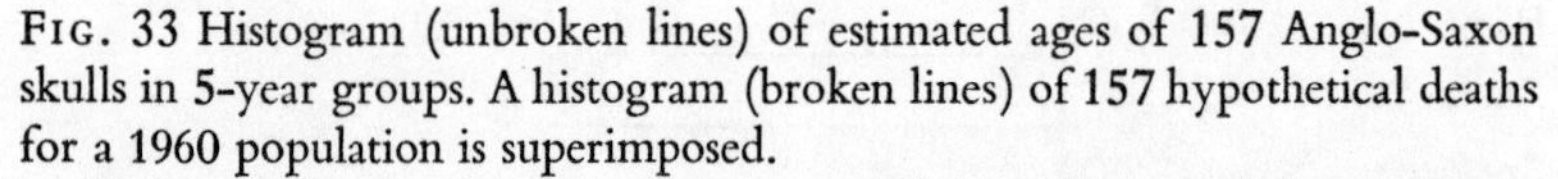

FIG. 33 Histogram (unbroken lines) of estimated ages of 157 Anglo-Saxon skulls in 5-year groups. A histogram (broken lines) of 157 hypothetical deaths for a 1960 population is superimposed.

After Comfort, A., *The Process of Ageing* (Weidenfeld & Nicolson, 1965).

and compares these with the age distribution at death of modern man. Very few Anglo Saxons lived to be over 40, an age when the death curve is only beginning to rise more steeply in modern man. These striking changes have been achieved by the methods already described: by reducing infections and deficiency diseases which would kill, or shorten life, at any age; by better systems of public health and by attacking the three captains of death in later life—coronary disease, cancer and bronchitis. If medical advance continues *without* any real breakthrough in the understanding of the theory of the ageing process Comfort believes that the commonest age of dying will shift from 75 to 85 and that not many more of us than the present one in 100 will reach 90 and not many more than the present one in 1,000 will achieve a century. The killers of today will have been removed only 'to uncover the next layer of the onion'—in short we shall be dying of diseases that are uncommon today. The prospect then, of living to 200 fitted with a plastic heart is a remote (and rather dreadful) possibility, for even though the heart, or liver, or kidneys, might in time be replaced, other parts of the body will almost certainly break down like the wiring in an old house. What would be useful and pleasant would be as Comfort says, to have the part of life which lies between 20 and 50 extended even by a small percentage, for it is between these years that vigour, health and creativity are greatest in most people. If this could be done without increasing the life span an almost square survival curve would be produced when people would remain young in mind and body for long periods and then die suddenly like the people in *Lost Horizon.*

THEORIES IN AGEING

Ageing seems to be caused by a gradual departure of the body from a regular pattern. It is, as Comfort suggests, an escape from co-ordination coupled with the arrears of processes which once contributed to fitness but are now out-of-gear and running free. The mechanism of ageing is still not understood in man although something is being learned about the process in animals. For example, the life span of some kinds of animals can be lengthened by altering their genetic make-up, the temperature at which they live, or their diet. Experiments with vinegar flies (which have a very short life cycle so the results of experiments can be assessed quickly) show that if the long-lived flies from successive generations are picked out and used to breed from, the life span does not alter but if flies from such a stock are crossed with those from another, the offspring can have twice the life span of the parents (see Fig. 34). The flies from these crosses exhibit what is called 'hybrid vigour'—that is they are large, live a long time, grow rapidly and are very fertile. While it is obvious that these flies have mixed genes from both parent stocks it is not known why such a combination of genes gives vigour.

The effects of temperature on life span have been shown on water fleas. Fleas kept at 8°C live longer than those kept at 18°C but are the same size. Fleas kept at 28°C have shorter lives and are smaller. If they are starved at any temperature for three-quarters of their life and then

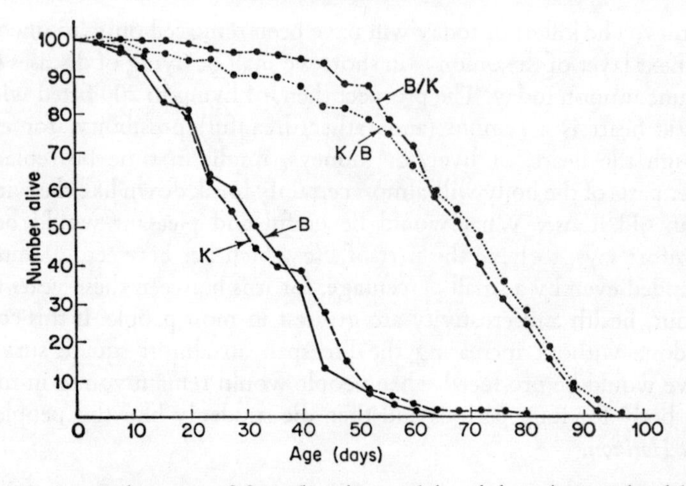

FIG. 34 Survival curves of fruit flies (*Drosophila subobscura*). K. inbred line; B, another inbred line; K/B and B/K, the reciprocal hybrids between them. The hybrids have a higher survival rate.

Clarke and Maynard Smith, 1965, taken from Comfort, A., The Biology of Old Age. New Biology No. 18 (*Penguin Books*, 1955).

suddenly given plenty of food they live longer than continually well-fed fleas. This bears out the old saw 'a short life but a gay one'.

Prolonged deep cooling by refrigerated blankets slows down the working of the human body until its activity becomes practically nil at a few degrees above 0°C. Two centuries ago John Hunter, a famous surgeon, had the idea that 'action and waste' would cease on freezing a person. On thawing the body (he thought) activity would start again and life would be lengthened. This idea has been resurrected recently in America. Hunter experimented with a pair of carps in 1766 but failed. He wrote:

> I thought that if a man would give up the last ten years of his life to this kind of oblivion and inaction, it might be prolonged to a thousand years; and by getting himself thawed every hundred years, he might learn what had happened during his frozen condition. Like other schemes I thought I should make my fortune of it; but this experiment undeceived me.

Rats kept on a normal but low calorie diet for about 1,000 days, the normal age of death by old age, and then given a normal diet live for about twice as long as well-fed rats, that is about 2,000 days. In the second phase of their life they grow to full size, become mature, and then old. And all this after a lifetime's wait. If, however, rats are starved *after* puberty their life is lengthened slightly but only because the reduced feeding prevented obesity. The long life of starved rats, therefore, is probably the result of prolonging childhood; their 'programme of life' seems to have been slowed down before the critical age of puberty. Starvation after puberty is not effective in increasing the life span for after about 1,200 days half the rats were unable to resume growth. Perhaps over-feeding of children, giving quick growth and early maturity is incompatible with long life, but there is no evidence, as yet, for this.

For clues to longevity in man it may be worth while too to study special populations such as those of the Caucasus republic of Georgia in the southern U.S.S.R. many of which are said to contain very old people. In fact in the last census this particular republic, whose population is about the size of Croydon's, claimed to have 2,000 people over the age of 100. And some of these were well past the century. These long-lived populations have been studied intensively by doctors and scientists from the Soviet Institute of Gerontology, who seem satisfied that the claims of the old people about their ages are real, despite the lack of support of birth certificates. Some of the old men (the men live longer than the women) can remember clearly incidents from the past such as the death in a duel of Pushkin in 1837 (although it is odd that peasants have even heard of Pushkin). Checks with parish registers however showed that the majority were telling the truth.

Detailed study and questioning of the centenarians reveal that they have always lived extremely regular lives. They get up early, eat three or four meals a day, always with plenty of highly spiced fruit and vegetables, and retire early at a regular time. Their way of life gives them plenty of exercise away from urban strain in a good climate at a high altitude. Most of them were born under primitive conditions. Nearly all are intelligent, happily married and retain their sexual powers until past 100. Most of them have children and stable and strong family ties. While drunkenness is rare, wine is commonly drunk in large quantities and smoking is moderate. All this seems an excellent recipe for a long and good life for anyone. Doubtless genetical studies will reveal that such long-lived people are inbred and carry genes promoting longevity. The sceptic, however, will insist rigidly upon documentary evidence of old age. Whenever this has been produced very few people indeed are shown to have exceeded 100. The oldest man in Britain in 1964 was 108. Very much higher claims than this are meagrely supported by documentary evidence and are probably untrue.

Three major theories to explain ageing in man and animals have been put forward. One is that the rate of growing old is controlled by the loss of certain vital time-keeping cells such as brain cells but as yet experimental damage to the brain of animals has not resulted in ageing. Another is that it results from dividing cells producing worse and worse copies of themselves. The third is very old and is based on the idea that signs of age arise when the secretion from the sex glands begins to wane but the use of sex hormones to rejuvenate the old has failed.

These speculations about the mechanism of ageing, although interesting, have not brought us any nearer to solving the problems of the process. But they do suggest pathways for new experiments that may provide a clue to giving longer life. As Alex Comfort says, 'if we destroy more hypotheses than we demonstrate, this is a subject which can well stand such treatment in contrast to the speculation which has gone before'.

SPARE-PART SURGERY

Fig. 35 shows clearly what can be done already in the way of mechanical healing, but what of the future? It is clear that all the organs of the body do not wear out simultaneously so that if a heart or kidney begins to fail, life might be prolonged by grafting in another from a bank stocked with organs from donors who have died. The trouble is that in the long run such grafts are not accepted by the body and they stop working. The tissues of each individual have a unique make-up which recognizes foreign (i.e. grafted) tissues and tends to reject them, even though the tissue is that of a life-saving heart or kidney. The body responds to the antigens of the graft by producing antibodies which destroy the cells of the graft and so reject it. It may be that these antibodies are found in certain white blood cells which are made in the lymph nodes of the host. The antigens, the substances that call forth the antibodies, are probably produced directly by the genes. Animals that are very alike genetically (closely inbred strains of rats and mice) and human identical twins (genetically identical) can successfully take skin grafts from each other. If the antibody-producing tissue could be destroyed it would be powerless to reject a 'foreign' graft. And if then the antibody-producing tissues were replaced by cells from another individual's organ, grafts might be accepted from this individual. This in fact happens in mice, whose lymphoid tissue has been destroyed by

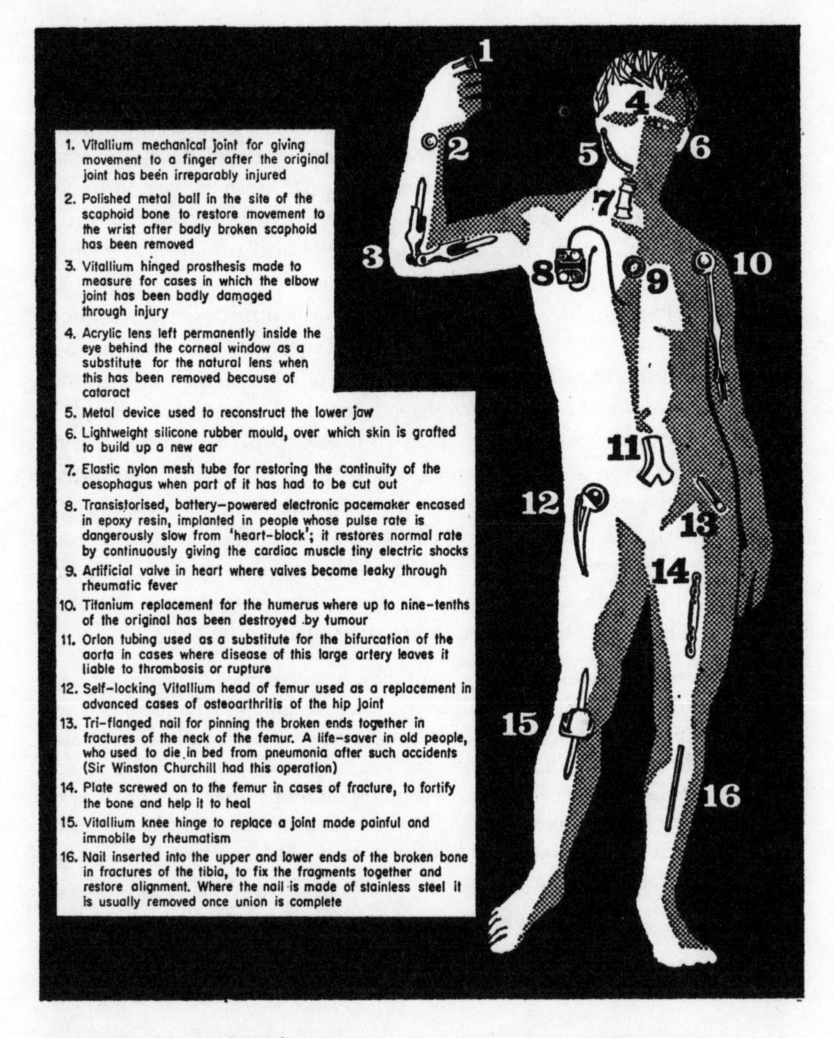

FIG. 35 Mechanical Healing
After *Sunday Times*, 6 December 1964.

X-rays and replaced by cells from another mouse. After recolonization of the lymphoid tissue by the new cells, organ grafts could be made with tissues from the mouse whose cells had been used to recolonize the lymphoid tissue. When a human is mortally ill from, for example, kidney disease it is possible to graft in a foreign kidney after X-ray treatment, but the graft does not usually last long. For mass replacement of faulty hearts or kidneys there would not be enough human donors even if the real obstacle to spare part surgery, the uniqueness of the body, could be overcome. Moreover it is possible that a black market traffic in spare parts might develop in time! More practical possibilities are: to develop mechanical hearts with valves and pacemakers (many of these are now in use), livers, voice boxes, bile ducts, bladder-controlling devices and other mechanical organs and parts of organs.

It has been suggested that pure strains of non-human species might be bred (e.g. monkeys) to produce material for spare parts. Simultaneously antigens, which would condition a human body to accept non-human organs, would need to be developed. Even if these ideas could be carried through many people would object to being powered by a monkey's or even another man's heart. Indeed the idea of the individuality of the body is strong enough for many of us to reject organ transplants in preference to mechanical devices. These, in any case, might well be developed much more quickly than grafting techniques.

In the not so distant future, then, it is possible that a man will be alive who is semi-artificial, and, moreover, glad to be alive.

He will wear eyeglasses, of course, and may have a cornea transplanted from someone no longer living; he will have a hearing aid and false teeth and will walk with a cane; his aorta—the body's main artery—will be a knitted Dacron tube, each heart beat will be sparked by a battery carried either under his skin or in a pouch slung over his chest, depending on his preference and that of his surgeon, daily he will take an insulin injection and twice weekly he will spend ten hours attached to an artificial kidney machine. He may be on the waiting list for a transplanted kidney.*

PREDICTION OF HUMAN ABILITY

The two previous sections have considered the possibility of extending life in the future. Probably of greater importance for human happiness and fulfillment, is the development of more precise techniques for determining human ability and skills so that latent talent does not go down the drain. J. B. S. Haldane considered that the recipe for happiness was doing a job which was difficult but not too difficult. Aptitude tests designed by psychologists *and* physiologists, especially

* Schmeck—see Bibliography.

brain physiologists, to determine variations in the functioning of the body and in personality may have important consequences in getting people into the right jobs, or starting them on the right career. For example, special physiological tests to discover supernormal vision may pick out the one person who might become a really superb dentist or lens or tool maker.

Perhaps very few talented people are missed nowadays when they apply for admission to higher education, but it is highly likely that Darwin's genius (and perhaps the genius of Faraday, Bunyan, Chopin and the painter Turner) would not have been recognized by the most penetrating and sensitive admissions tutor. Darwin had a poor record at school and at the university. He chopped and changed, dropping first medicine and then the church as possible professions and managed at last to scrape an ordinary degree at Cambridge. In these highly competitive days he would be as Liam Hudson comments, 'just a nice young man with a good collection of beetles' and probably two shaky 'A' levels.

It is important for the future of any country to recognize early, and foster by all the means it can, original and creative people who are not necessarily of the highest I.Q. but who seem to have a high degree of persistence, who seem able to channel their ability in an obsessive, even an erratic single-minded way. Usually, so it appears, such minds differ from 'non-creative' minds by being more 'open' and speculative, 'less hidebound in attitude and belief and exceptionally self-reliant'. Einstein, for example, decided for himself which parts of his education were worthwhile and which were not.

At present we do not know much about the external conditions which lead one man to become a scientific or artistic genius, except that he is very likely to have had parents of above average intelligence. We need to know what predisposes a man to follow a particular bent, how to lead him in the right direction and how not to crush his talent under the weight of examinations or a rigid curriculum. Perhaps the basic clue in the detection of these people lies not in I.Q. testing alone but in the sphere of specially designed personality tests.

The picking out of genius or of excellence is certainly one job of the educational system but naturally its greatest task is to promote the general betterment of mankind by initiating young people into worthwhile activities.

To go some way to achieve this we need to recognize that human variety is the strength of society. We have seen that all men are not created equal and are genetically unique. For education this means the development of more discriminating techniques to discover the inborn talent of young people coupled with the skill to guide them into a variety of educational systems and occupations. We need to learn to discover the visualizer as well as the verbalizer, how to spot creativity and imagination as well as quantitative, analytical ability. We need to know whether, for example, musical, scientific and artistic abilities are correlated or independant and if they are, to what degree this is so. We are still in the Stone Age as far as these important matters are concerned. Even so we have moved a little. At one time selection was based on hearsay and subjective evidence—choral scholars are an extreme example of this. But even now the failure at 11+ is still a failure on a test concerned with a *single* value scale. Before we condemn this too quickly, however, we must remember with gratitude that there are many people today of humble origin who are now famous in science, law, medicine, politics and literature who, as Sir Cyril Burt has reminded us, would never have reached the top if they had not taken an intelligence test, the 'scholarship', at 11+ which revealed potentialities that might have gone unnoticed and undeveloped.

Clearly selection is required and can and does happen at 5, 8, 11, 13, 18 and 21 (for Ph. D entry) but perhaps what is necessary right up the age scale is *refined selection with reselection*; the facility to retest and correct errors. This is of supreme importance if human inborn potentiality is not to be wasted.

Part 4: The Crisis of Numbers

It is useful to look at man's problems in relation to where he lives by treating him as an animal subject to the same checks and balances as other animals. If he over-eats his food supply the harsh checks of starvation and disease start to operate. Thus, like any other animal, man must be able to reach a state of equilibrium with his limited environmental resources. Quite unlike any other animal species which has no chance to improve its lot, he is largely in control of his environment. Three centuries ago man almost everywhere lived under the shadow of Darwin's struggle for existence; fever and famine especially and sometimes war cut his numbers drastically. From the eighteenth century onwards mortality declined and population increased in Europe principally due to a reduction in deaths from tuberculosis, typhoid, cholera, dysentry and diarrhoea and smallpox, a reduction not due to medical advance but to improvements in living standards. These 'old' diseases have now been controlled in rich countries and even in developing countries child mortality is not so great as it was because all the medical knowledge of the last two centuries is available to them at once. It is in these countries that famine is beginning to exert a control on population.

The microbe has always been a greater killer than the sword, bullet or atom bomb, but war with germs might, if any country was inhuman enough to use them, bring back plagues as rampant as the Black Death.

As population increases man's habitat is slowly but steadily being ruined. If anyone wishes to see what man has done to his habitat he should go to Tunisia; once a granary of Rome it is now through neglect and decay of the soil a semi-desert. The Thames too, once clean and filled with life, including the salmon and oyster, is now another kind of desert where the blood worm, a creature adapted to oxygen-poor mud, thrives, a last outpost of life.

Growing population means growing cities, for these are the necessary money-spinners for any economy. Not only does their growth eat up land for aerodromes, roads and reservoirs but their sheer size brings with it special health problems and all the frustrations of travel in and out of the city. Population growth is at the root of most of man's ills today; it is not only self-destroying in an environmental sense, but it is the basic cause of mass poverty, mass ill-health, mass-unemployment and mass illiteracy. Population must be brought down to a manageable level.

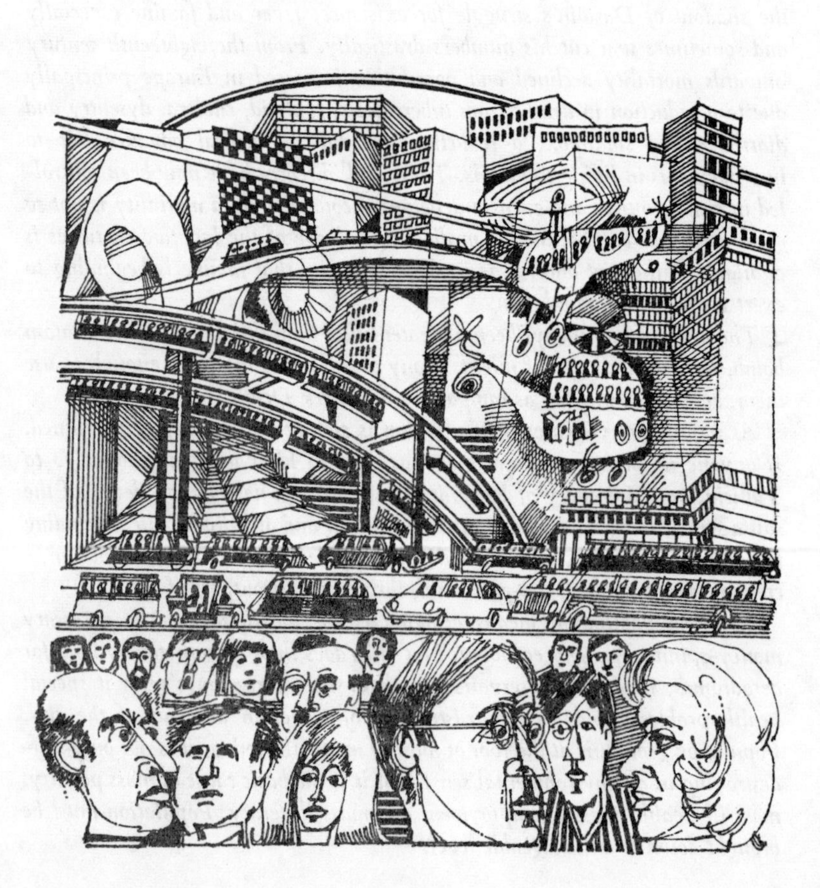

20: Population Problems

IT was hinted in Chapter 16 that once it was possible to control the diseases of poor countries, and this is now possible given money and education, then over-population and starvation (and thirst) would result. In the past, and even now in some countries, if man multiplied too fast for his food supplies he, like any other animal, starved or lived in continual danger of being attacked by disease. Sometimes in the long past shortage of food forced him to migrate in the same way as a shortage of lemmings and ptarmigan in the Northern Tundra causes the Arctic Fox to migrate southwards in large numbers to escape starvation and disease. Table 17 shows some of the checks to population which have taken place in the past. Although, as the table shows, war and famine often whittled down populations, infectious disease was the main check to population growth right up to the nineteenth century. This check still operates in many poor countries, but in Europe the brake put on population growth by disease has been taken off slowly over the last three centuries. For example, in about 1650, the world's population stood at approximately 450 million, of which about one quarter lived in Europe and about six million in the British Isles. Ever since that time, the population of Europe has increased but why

TABLE 17

Type	*Some population checks*	
Disease	Black death, 1348–50	20 million people died in Europe out of a population of 85 million
	Cholera, 1831–66, Britain	200,000 died
	Influenza pandemic, 1918	100 million people died
	Malaria, 1953, India	800,000 died
Famine	India, 1769	10 million people died
	Ireland, 1846–7	Between two and three million died in potato famine
	China, 1877–78	9 million died
War	First Punic War, 255 B.C.	95,000 died when Roman Fleet was wrecked
	Mexico, 1519–48	18½ million people exterminated by Spaniards out of a population of 25 million
	World War I, 1914–18	8 million dead and 37 million casualties
	World War II, 1939–45	22 million dead. Nine million died in concentration camps including 6 million Jews
Natural disasters	Bengal earthquake, 1943	1½ million died
	Skopje earthquake, 1963	1,070 died. 3,300 injured
Man-made disasters	Great fire of London, 1666	6 died
	Bombing of Dresden, 1945	about 35,000 died
	Nuclear bomb, Hiroshima, 1945	About 80,000 died

it should have started to do so then is a mystery, for disease was not under control. Perhaps the new crop plants, like the potato from the New World, gave Europeans a better diet. Certainly after 1665 (the year of the Great Plague) plague vanished from England and Europe. This disease was a savage killer. It slaughtered the Philistines in 320 B.C. and was responsible for two of Europe's greatest population checks—

the plague of Justinian's reign (A.D. 542) and the black death of 1348. Between 1348 and 1350 at least a quarter of the European population was wiped out by the black death and by 1400 a third of the population had died. Florence lost half its population of 90,000 people and Sienna about 27,000 out of its 42,000. From 1348 to 1374 the total population of England dropped from 3·8 to 2·1 millions. In Western and Central Europe as a whole the death roll was so great that it took nearly two centuries for the population level of 1348 to be regained (see Fig. 36). From the thirteenth century to the year of the great Plague (when a tenth of London's population of half a million died) Europe was riddled with disease; not only plague (a form of black death) but syphilis,

IMPACT ON POPULATION FROM RECURRENT PLAGUES IN EUROPE

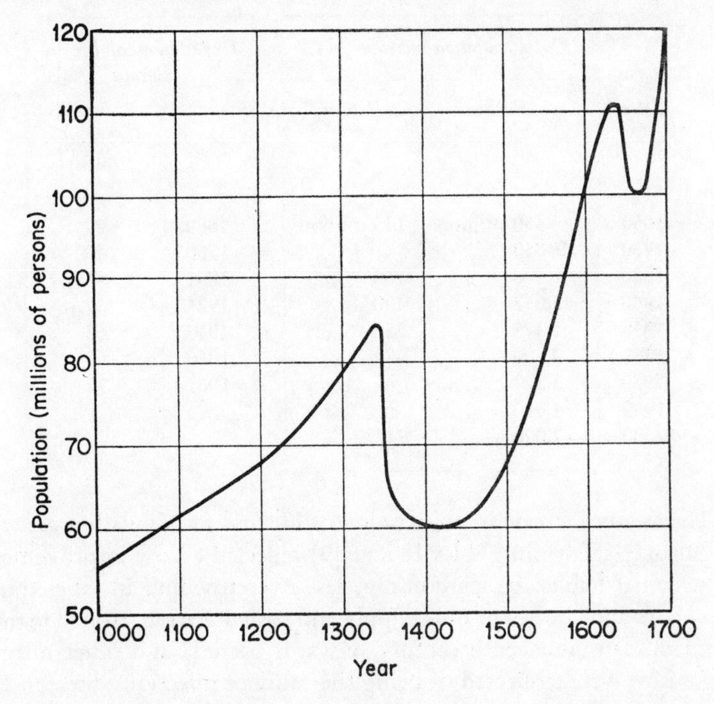

FIG. 36 For more than 300 years after 1347 the plagues checked the normal rise in population; sometimes, as in the 14th and 17th centuries, they resulted in sharp reductions. The figures shown on this chart derive from estimates by students of population; actual data for the period are scarce.

After Langer, W. L., 'The Black Death', *Scientific American*, February 1964.

typhus and a deadly form of influenza known as the 'English sweat'. But what caused plague to vanish in England and Europe? Improved diet has been mentioned and this might have increased man's resistance to disease. Black rats which carry the fleas that transmit the disease to men were gradually exterminated by brown rats in the early eighteenth century and this ecological change might have stemmed the disease. But this is unlikely for it had begun to wane decades before this. Much more likely is the explanation of a mild strain of the disease-producing bacteria replacing the virulent strain in the flea. We now know this happens with many viruses, producing severe and mild attacks of disease.

TABLE 18

	Population increase	
Date	*World*	*Europe*
1650	450 million	145 million
1750	650–850 ,,	?
1850	1,000 ,,	265 ,,
1900	2,057 ,,	400 ,,
1950	2,475 ,,	550 ,,
1965	3,300 ,,	639 ,,
1970	3,570 ,,	?
1980	4,000 ,,	?
2000	7,000 ,,	947? ,,

TABLE 19

Expectation of life in Europe	
Date	*Expectation (men)*
1650	27
1750	40
1881	45
1921	57
1940	63
1950	67
1967	72

The steady increase in expectation of life and in population growth in Europe (shown in Tables 18 and 19) reflected a slow improvement in personal habits, better housing, less overcrowding in cities, pure food laws and control of milk supply and pasteurization. Added to this, from the late nineteenth century onwards bacteria and other micro-organisms were suspected of being the cause of infectious diseases. By 1900 most of the important disease bacteria had been discovered as well as the ways in which they cause infection. Fig. 37 shows dramatically how, for example, typhoid deaths were reduced between 1871 and 1931 as a direct result of improved sanitation, anti-typhoid inoculation and the break-up of over-crowding in cities by better systems of

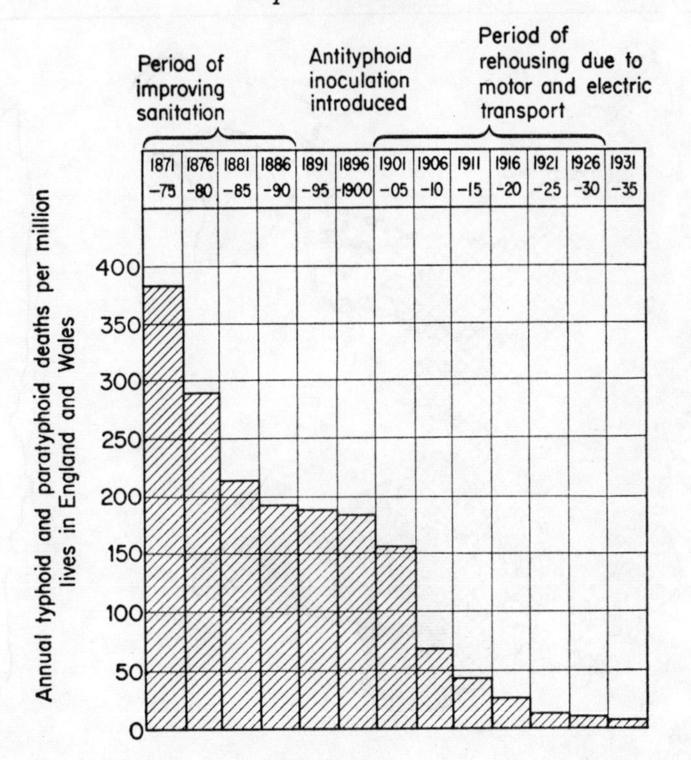

FIG. 37 The decline of typhoid deaths from various factors. From *Science Past and Present*, Sherwood-Taylor F. Mercury Books, 1962.

transport which allowed people to live in healthier surroundings at reasonable distances from their work. It is important to stress that improved *social* conditions were the real causes in the decline of mortality not advances in medical care.

In Europe and North America at the present time infectious disease presents no problem. Aseptic surgery, vaccination, immunization, the use of antibiotics and the simple application of hygienic measures have conquered the 'old' diseases of typhoid, tuberculosis, cholera, scarlet fever, dysentery, plague and a host of others. To develop these techniques of disease control has taken about three centuries and this accumulated wisdom is available at once to underdeveloped countries, with resulting population problems. Fig. 38 shows the relationship between fertility and death in the world and it will be seen, for example, that a large part of Africa has a high fertility rate and a high death rate

POPULATION GROWTH IN DIFFERENT REGIONS

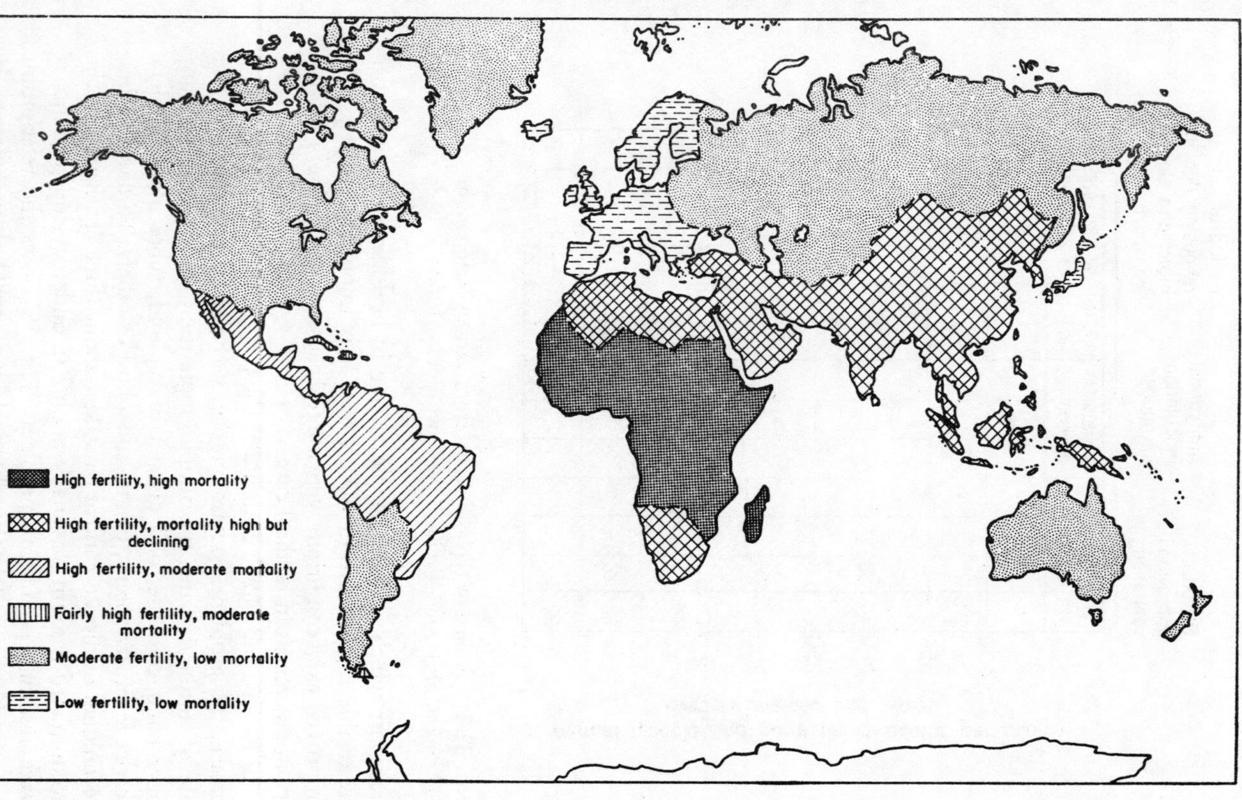

FIG. 38
After *Man and his Environment* (B.B.C. publication, 1965).

in contrast to Europe's low fertility and low death rates. (Low birth rates are somehow related to high standards of living.) One example shows dramatically how modern medicine can cause a population explosion in areas of high fertility but high infant mortality such as South America. Back in 1945 Georgetown in British Guiana had a very high rate of deaths among babies—about 250 per thousand—and insect-borne diseases were said to be the cause. So DDT was sprayed periodically from aircraft over an area of 10 square miles including Georgetown. The spraying went on from 1945 to 1948 with quick and dramatic results. Infant deaths dropped from 250 to 67 per thousand and the birth rate rose. This was a mixed blessing because the economy of British Guiana can just support its population above the brink of starvation and the increase in numbers could have only one result—starvation and misery. This case history, and many others could be quoted, shows what might happen to areas such as Africa and Asia if disease was controlled without a reduction in birth rates.

Nearer home the Irish famine of 1846–8 illustrates clearly the link between food and population numbers. The Irish depended on the potato for food, and its failure due to blight brought famine and disease to the eight million people. In the course of 1846 and 1847 between two and three million Irish died and over a million emigrated. Now the Irish population stands at about three millions—the result of a change in marriage customs to cope with poverty. Fear of the return of famine encouraged late marriage or celibacy in both sexes.

CONVENTIONAL FOOD PRODUCTION

The example of Georgetown illustrates clearly the effect of infectious disease (and the prevention of infectious disease by modern medicine) on the birth rate, while the Irish example demonstrates the link between food and population size. Studies with both animals and plants show that population size is regulated by food supply. Man is no exception to this but his skill and inventive capacity turned itself to improved methods of food production once urban population started to increase. In short, population growth itself changed agriculture from exploitive to progressive methods; the palaeolithic hunter needed about 10 square kilometres to feed each of his family; the neolithic herdsman needed 0·1 square kilometres, or 10 hectares; the medieval peasant raised his cereals on two thirds of a hectare of ploughland; an

Indian needs about one fifth of a hectare on which to produce his rice while a Japanese needs one sixteenth of a hectare (640 square metres).*

In Western Europe the change from haphazard to progressive agriculture came in the Middle Ages in the neighbourhood of the rich urban communities of Flanders. The Flemish farmers responded to the demands for their products and intensified their farming by buying garbage and human excrement from the cities to manure their fields. In the seventeenth century wood and peat ash (from Holland) were brought in to improve the fields near to the towns. The intensive winter feeding of livestock made possible by including clovers and turnips in the rotation helped to conserve the basic elements of fertility—nitrogen and phosphorus—by returning the cattle muck to the land. And the clover besides supplying food to cattle returned nitrogen to the soil from the air. A system of conserving fertility based on the Flemish system soon spread to England.

At the other side of the world the Japanese raised the yield of husked rice from about 15 cwt per acre in the period 1878–82 to 29 cwt in 1956–60. Rice cultivation depends on controlling water on the land; virtually no losses by erosion occur while fertility increases slowly from silt brought in from outside by the water. Human excrement and compost added to this and lately fertilizers improved the soil. And the Japanese farmer tilling his tiny unit of land on the side of a mountain has the added stimulus of developing cities.

Can the population problem be solved by even more intense farming? It looks as if we can go a long way to producing more food on conventional lines. For example, it is claimed that grassland in the hot, wet, tropics, when fertilized, can produce three times as much as the best temperate grassland. It is also capable of growing three cereal or leguminous crops each year when fertilizers are added. These wet tropics (Africa, Latin America and South-east Asia) probably have 510 million hectares of potential agricultural land which are equivalent to 1530 million hectares of temperate land. Not counting these wet tropics the world has the equivalent of 6,660 million hectares of good or potentially good agricultural land. In order to support a world population of 45,000 million people (possible in 150 years time) and to give them a diet containing meat and dairy products in plenty, 8,200 million hectares are necessary. To produce this land is easier said than done but the bare recipe is this; the need to break down resistance to scientific methods of food production by peasant conservatism; the

* Colin Clark (see Bibliography).

need to look after soil; the need to know how to use modern fertilizers properly; the need to understand biological food chains so that toxic chemicals do not cause gross disturbances in the 'balance' of nature; the need to improve seed and stock by breeding experiments; the need to harvest the riches of the sea; the need to irrigate and control flooding and the need to look for novel ways of producing food.

Some of these basic requirements are discussed in the next section but one very important development necessary to better the food production of the developing countries appears to be a paradox: it is

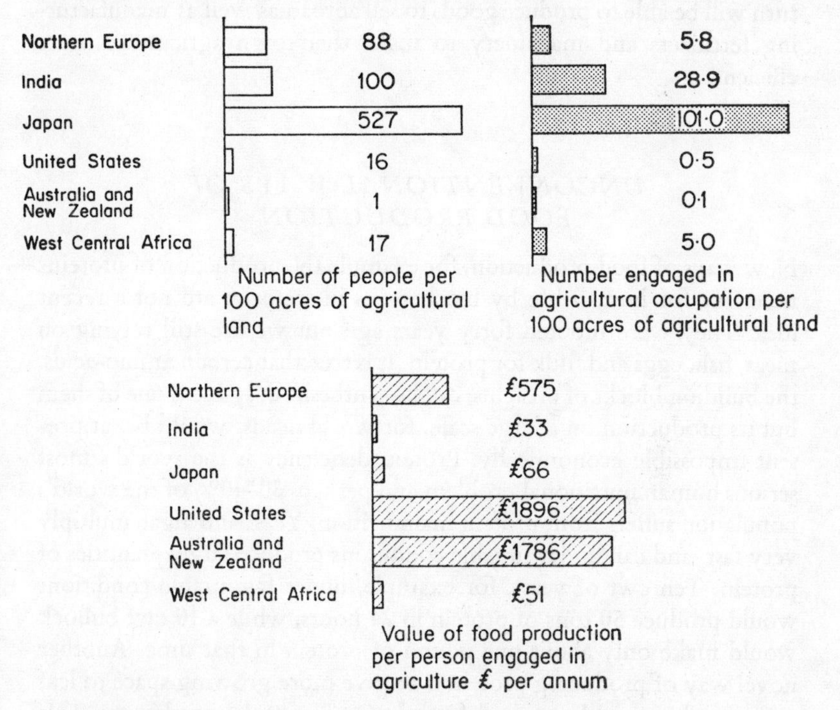

FIG. 39 Agricultural output of countries representing the 'two camps' in world agriculture. In developing areas such as India and West Africa about 30% of the population are engaged in producing food, while in the West far fewer workers are needed to produce much larger quantities of food. Supported by a massive industrial output, Japan, with her enormous density of population, has already raised her very modest level of agricultural productivity out of the poorer group of countries.

After Hutchinson, J. B. and Wilman, D. 'The Strategy of Food Production', From Harrison, G. A., *et al., Human Biology*. Oxford 1955.

that many of the world's farmers must move out of agriculture into cities. In prosperous areas of the world like northern Europe, the United States and Australia and New Zealand as Fig. 39 shows, very few people work on the land but the efficiency of these in producing food is enormous. The reasons for this are efficient fertilizer industries and good mechanization. And this is based on wealth produced by the industrial strength of their cities in which most people work. The prime need of developing countries like India is to acquire better balanced economies in which the land feeds urban populations who in turn will be able to produce goods to sell abroad as well as manufacturing fertilizers and machinery to make their own agriculture more efficient.

UNCONVENTIONAL WAYS OF FOOD PRODUCTION

New ways of food production, for example the production of proteins necessary for human life by biochemists in factories are not a recent idea. They were mooted forty years ago but we are still relying on meat, fish, eggs and milk for protein. It is true that certain amino-acids, the building blocks of proteins, can be synthesized; lysine is one of them but its production on a large scale, for world needs, would be, at present impossible economically. Protein deficiency is the world's most serious human nutritional problem and perhaps 30–40% of the world's population suffers from protein malnutrition. Yeast and algae multiply very fast, and can under suitable conditions produce large quantities of protein. Ten cwt of yeast, for example, under favourable conditions would produce 50 tons of protein in 24 hours, while a 10 cwt bullock would make only about one pound of protein in that time. Another novel way of producing protein is to give more growing space to leaf protein which can be pressed from leaves by special machinery. This could be canned or kept in refrigerators. Mr N. W. Pirie who has developed the leaf protein idea claims that a machine fed with moist leaf containing 20% protein at one ton per hour for 8 hours on 300 days a year would provide about 10% of the protein requirements of 30,000 people. The capital cost of such a machine would be about £2,500 and it could be used to extract protein from waste leaves of plants such as sweet potato, jute, sugar beet, sugar cane. It would be possible for a mixture of maize as a calorie source and leaf protein as a protein source to provide an adequate diet but it would need to be made to satisfy the human palate as well as to look attractive for the table. Also, before the leaf protein idea could go forward on a large scale, experiments need to be made to determine whether leaf proteins are as good as beefsteak when fed with other components of a normal diet and if so what the cost of producing them on a large scale would be. Until then it appears that the inefficient farmer co-operating with inefficient natural processes can still produce these essentials of diet more cheaply than the factory, and observations in human behaviour all point to the fact that most of us want to eat more of what we already eat.

RICHES OF THE DEEP

Harvesting the sea has already been mentioned. The sea is potentially a rich source of protein yet only about 10% of the world's animal protein and 1% of the world's total food supplies come from the sea. The main reason for this is our ignorance of 'oceanography' or 'inner space' as the Americans have called it. Indeed we probably know more about the surface of the moon than the seas on our own planet. Our knowledge of fish movement is very sketchy and the actual techniques of fishing are primitive. Yet the sea covers over 70% of the earth's surface; 139 million square miles with an average depth of 12,500 feet.

Clearly there is enormous scope for developing fishing as a major industry, as well as raising fish in tanks inland. Fish cultivation in ponds could, under good conditions, yield up to two tons per hectare per year. Plankton, the drifting tiny animal and plant life of the oceans, which forms the base of food chains in the sea, are particularly rich where upward currents bring phosphates to the sea surface and provide rich grazing for fish. Judging by the weight of plankton formed, and this depends on the rate of photosynthesis and the upwelling of salts, certain parts of the seas are as good as ordinary farm land. In time our descendants may be able to fertilize the oceans as well as the soil and reap a rich reward. One idea from America is to install a complete atomic power plant at the bottom of the sea. Its heat output would cause an upwelling of salt-rich sea from the bottom. At the surface the salt would increase the rate of plankton production. Fish would multiply feeding on plankton and suction devices beneath ships could harvest the fish crop economically.

Peering into the far distant future when man's skill and wealth might make it unnecessary for him to work hard physically, large quantities of food will perhaps not be needed. Then too, it may be that some of our descendants will be living on other planets with artificial supplies of oxygen and water, or even living under the sea and kept alive by synthetic foods.

WATER

Whatever happens to our descendants in the distant future the water demands of advancing technology, irrigation and hygiene in the developing countries will probably double by 1985, so that not only will much of the world be hungrier, it will be thirstier too. As more roads and factories are built, more steel is produced and more wheat is grown, so more water will be necessary because it takes about 15,000 gallons to produce a ton of dry cement, 65,000 gallons to produce a ton of steel, and much more than this to produce a ton of wheat. Added to this is the extra need of water for washing, drinking and cooking as the population grows. In countries with western standards of living the amount needed per head is about 50 gallons per day. In developing countries about three gallons per head are used at present (see Chapter 23).

Water resources are fixed and there is no possibility of making any more. Indeed there is plenty of it, but most of it is unusable and what is usable comes down as rain in a patchy and capricious way so as to leave huge areas of the world parched. In fact about 97% of all water is in the sea and about 2% is locked up in glaciers and ice caps leaving only about 1% of this undervalued commodity in circulation as fresh, usable water. Thousands of square miles in North and South Africa, the Middle East, India, China, do not see rain for over a year. Even in Britain which is very small and has an estimated rainfall of 80 billion gallons a day, rainfall is very unevenly distributed, too much of it falling in the north-west (see Fig. 40) and not enough in the south-east where population is densest.

Water shortage is becoming so large a problem that in 1962 President Kennedy said: 'There is no scientific breakthrough, including the trip to the moon, that will mean more to the country which first is able to bring fresh water from salt water at a competitive rate, and all those people who live in deserts around the oceans of the world will look to the nation which first makes this significant breakthrough. . . .'

Getting fresh water from the sea on a large scale is being tackled now in many ways. These vary from the age-old African method of using certain soils to remove the salt from brackish water to multi-stage flash distillation. In this process heated sea water is passed through a series of partial vacuums, thirty or forty of them and at each stage some water instantly vaporizes or 'flashes' into steam which condenses out as fresh water. Wasted heat from nuclear and conventional power stations, and

DISTRIBUTION OF RAINFALL AND POPULATION

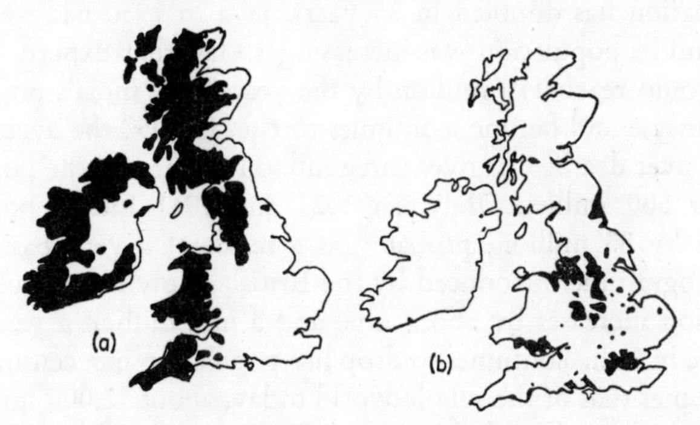

FIG. 40 (*a*) Shaded area represents precipitation over 40 in per annum. (*b*) Shaded area represents a population density of over 500 per sq. mile.
After Ovington, J. D., 'Vegetation and water conservation', *New Biology*, 25.

perhaps solar heat in arid regions, could fire the plant. British engineers have designed a dual purpose nuclear plant to produce electricity at 400 megawatts a day as well as converting sea water to fresh water at a rate of 60 million gallons a day. Such a power/water plant would cost about 65 million pounds, a great deal of money for a developing community who no doubt would be able to use the water easily but would find difficulty, at first, in harnessing the electricity. However the balance between electricity and water need could be adjusted to local needs. While the future of turning sea water to fresh water by complex machines looks bright it must not be forgotten that icebergs are almost pure water. A big iceberg according to some Americans could be worth 200 million dollars at wholesale prices!

SOME PREDICTIONS ABOUT NUMBERS AND FOOD SUPPLIES

To end this chapter it may be useful to state some facts and predictions about population increase. The *Population Bulletin* for 1950 gave ten areas where man is breeding at a prodigious rate. Among these were Ceylon and Puerto Rico, where it was predicted that the population would double in 25 years if the 1950 rate of growth continued. Egypt, another danger point, has had a sixfold increase since 1850 and its population has doubled in 35 years. Java in 1950 had 44 million people and its population was increasing so fast that experts believed that it would reach 110 million by the year 2000. India's population, despite disease and famine, continues to rise. In 1950 the average gain of births over deaths was over three million a year and the population was over 500 million. Between 1921 and 1941 India's population increased by 83 million, probably as a result of a very basic public health programme introduced by the British. Now her population of 500 million increases by about nine and a half million a year. If the death rate in India continues to drop her population in a century could be four times that of the whole world today, about 12,000 millions.

All the countries listed above, and there are others like China with a quarter of the world's population, are having the old checks of disease and famine (see Table 17) slowly removed and there is bound to be a population build up. On a world scale, in ten years the population will have risen from 3,000 millions to around 4,000 millions. By A.D. 2000 it will be 7000 millions. To match the rise, in ten years we shall probably need, on a world basis a 50% increase in total food and a nearly 60% increase in animal food.

In poor areas like Africa and India even with a small rise in meat eating, animal protein will have to be boosted by 120%. By A.D. 2000 total food production will need to go up by 170% and animal foods by 200%. In underdeveloped countries the rise in animal protein will have to be something like 500%. There may well be no difficulties in producing the calories but there will be in producing the protein. As well as producing more food the population problem needs to be tackled from the other end—by birth control. It is improbable that the span of life will shorten and it is likely that the 25 to 50% of infant deaths in poor countries will be reduced fairly quickly.

21: Population Control

CRUDE methods of controlling fertility are old. In primitive societies lack of food and weakness due to long-standing illness may have applied a natural check. The Romans knew that hot baths reduced male fertility and modern experiments have confirmed this. Baths around 45°C and of a quarter of an hour's duration, taken over three days, will cut a man's sperm count to well below the number likely to be effective in fertilization. If this bathing drill is repeated every fortnight a fair degree of 'natural' sterility will be achieved. In Malaya Chinese girls eat pineapple shoots to stop them having babies, with apparent success! Table 20 shows some evidence of primitive and rather drastic and cruel ways of regulating family size. For instance, in populations which have to search for food and water over large areas it is very difficult for the mother to carry more than one child if she is to keep up with the rest of the group, and sometimes a baby is killed or left to die.

The rest of this chapter is devoted to modern methods of controlling fertility designed to quell the silent but sinister population explosion in such places as Puerto Rico and India. In these countries and in others like them where the mass of people are ill-educated and ignorant, and

TABLE 20

Primitive ways of regulating family size

X—recorded instances; /—no recorded instances)

Group	*Killing the baby*	*Abortion*	*Restriction on intercourse*
Food collectors and hunters			
Australian tribes	X	X	/
Tasmania	X	X	/
Bushmen	X	/	/
American Indians	X	X	X
Eskimos	X	X	/
Agriculture			
American Indians	X	X	X
Africa	X	X	X
Oceania	X	X	X

From Harrison G. A., *et al.*, *Human Biology*. Oxford 1955.

live under unhygienic conditions, conventional methods of birth control are quite unsuitable. An ideal contraceptive is one that need only be taken occasionally in a pill form and should have no effect other than preventing conception. From this idea 'the pill' was developed. It has been tried out by 130,000 women in this country and by women in Japan, Ceylon, Singapore, Hong Kong and India. In America about two million women are using pills of one sort or another as contraceptives.

HOW THE PILL WORKS

About ten variants of the pill exist but all contain a mixture of two synthetic steroid hormones, one of which is oestrogen which stops the egg from being released from the ovary in ovulation. This probably acts on the hypothalamus (an area at the base of the brain) to prevent it secreting a substance which enters the pituitary gland in the brain and causes it to release a hormone into the blood which plays a vital part in ovulation. By taking a daily dose of one pill ovulation is stopped and changes also are brought about in the lining of the womb which makes it antagonistic to sperms. It has been shown that faithful pill-taking gives almost complete protection against pregnancy and is much more efficient than other means of birth control. This is shown in Table 21.

TABLE 21

*Effectiveness of different methods of birth control**

Method	*Average pregnancy rate per 100 woman years†*
Douche	31
Safe period	24
Jelly alone	20
Withdrawal	18
Condom	14
Diaphragm (with or without jelly)	12
Enovid (the Pill)	1·2

* Modified from G. R. Venning, *British Medical Journal* (p. 899, *2*, 1961).

† A 'woman-year' is 12 monthly cycles of exposure to normal sexual experience. It includes certain 'personal' failures as well as 'method' failures and so the results of any one method will vary according to the intelligence and experience of the user.

Even though the pill is successful in preventing unwanted babies occasionally it can have some nasty side effects. Women have reported headache, dizziness, nausea, vomiting, or depression, after taking it. Gross obesity, liver trouble and a tendency to develop blood clots in the veins are also said to be aggravated by the pill. Perhaps more serious than these immediate effects are the possible long-term effects of pill taking. No one knows what will happen to women whose hormone balance has been upset for say 20 years—from 20 to 40 years of age. The pituitary gland is the master gland of the body, the so-called leader of the endocrine orchestra, for its chemical signals discharged into the blood regulate the activity of all the other glands in the body. If the balance between the pituitary and other glands is upset over a long period, serious disease like cancer of the womb or breast might result although there is no evidence for this at present. There is some indirect evidence however that women with a high *natural* level of oestrogen in the blood are more prone to cancer of the womb than those with normal levels. Since the pill contains artificial oestrogen and therefore raises the amount of this in the blood it might, over a long period of time, cause cancer. In about 15% of women, long-term pill taking may lead to coronary thrombosis which we have seen is primarily a disease of men. This may be because the pill causes an increase in sugars and fatty substances in the blood. Although a link between these substances and thrombosis is not proved some men who have suffered heart attacks have an excess of cholesterol and sugar in their blood.

Another group of hormone-containing contraceptive pills makes the lining of the womb inhospitable to the fertilized egg provided the pills are taken on the day following intercourse, but these experiments have only been done as yet on rats and mice and have not been applied to man. Indeed there may be the world of difference between the results of animal and human experiments using the same pills.

Nevertheless both types of pill described—those that suppress ovulation and those that stop implantation of the fertilized egg—are probably the most promising lines of research into chemical contraception. Although, as a first step, carefully controlled animal experiments are essential to find out as much as possible about what the pills do and what their side effects are, the only real way to discover their effect on women is to test them on a large scale. All the women should be volunteers who have been told of the risks involved. In Asia millions of women might be glad to test the pill or pills, for their short life is often made a misery by incessant child-bearing which in itself often causes death.

Women who cannot be relied upon to take a daily pill might possibly be able to have a monthly injection of a long-lasting progestin-oestrogen compound similar to the pill. Trials with this injection have been successful and after injections have been stopped normal menstrual cycles with the production of an egg returned in 6–13 weeks.

THE PLASTIC COIL

One of the startling facts about the populations of developing countries like Africa, Asia, and Latin America, is that half of the people are under 15 years of age. Such young populations are highly fertile, ignorant of birth control and have an urge to mate which is adapted to low conditions of survival and not to those of the twentieth century.

In a recent symposium Professor Sargent Florence pointed out that, for poor countries at least, the pill may not be the most acceptable, effective and cheapest answer to birth control for the mass of the people. The plastic coil, the latest birth control measure to be introduced, may be more suitable. The technical name for this is the intra-uterine contraceptive device, I.U.C.D. for short. Rather than interfering with fertilization it possibly stops implantation of the fertilized egg in the womb. Unlike the pill it does not affect the balance of hormones in the body and has no side effects but there is always the possibility of it coming out of position.

POLITICAL CONSIDERATIONS

While it may be plain to westerners that the teeming millions of Asia need help to reduce births, on reflection it might seem to some of them that the west could be trying to reduce the fertility of the whole Asian population for political ends. This might be particularly true when ill-thought out suggestions are made that contraceptive substances might be put in everyday foods like salt which could be used by the women in a house when they wanted it. Perhaps this might be simpler than pill taking but for one thing there is so much uneveness in salt taking that the idea is not really practical. Another more important reason for rejecting the idea is that such 'doctored' food might lead to powerful revolt. In any case the decision to have or not to have children and a family of a particular size for its own sake is a human right and a human freedom. Some form of cheap, simple, safe contraceptive will at least remove the fear of having an unwanted child in a mother who might be in no state to have another baby. Chemical contraceptives involve risk but the risk must be carefully weighed against fear of unwanted and possibly dangerous pregnancies.

MASS CONTROL

Birth control on a large scale in developing countries has got to be cheap for the countries are poor; if some simple means of checking fertility could be achieved then the standard of living could rise. Indeed the social problems of poverty, unemployment and illiteracy can only be met by introducing cheap methods of birth control so that population does not outstrip resources. If it does, nature will apply to man, as to any other animal, the crude and harsh checks of famine and disease.

Table 22 shows the acceptibility, effectiveness and economy of operation of different methods of birth control in the city of Taichung in Formosa. Information about birth control was spread by means of posters, by meetings with leaders to get the co-operation of the people and by letters to newly-married couples and parents with more than two children. In addition personal visits were made by specially trained midwives to every married woman between 20 and 39 years of age.

The results of the experiment were encouraging for about 40% of women between 20 and 39 years of age took up birth control in the first 13 months. A glance at Table 22 shows that the I.U.C.D. with its

TABLE 22

Comparison of the Different Birth Control Methods

Cap	*Washable sheath*	*Pill*	*Plastic coil*
A. Acceptability:			
Distasteful to many women In Formosa only 15% accepted without need of home visit. Preferred to other methods by 20%		Requires sustained habit but regularity makes easier In Formosa only 2% prefer to other methods	No need for sustained effort In Formosa 75% accepted without need of home visit. Preferred to other method by 78%
B. Effectiveness:			
High if instructions carried out		100%	In Formosa 20% expel or remove after six months
C. Economy:			
Materials—A cap once a year (8*s*.) + chemicals	2 a year approx. 2*s*. 3*d*. each	6*d*. each, 300 a year £7 10*s*.	2*d*. each not more than one a year
Labour—Doctor for prescribing; nurse for teaching	None	Qualified doctor to supervise	Qualified doctor for insertion
Special Premises—A clinic; one or two visits a year	None	None	A clinic visit once a year

From P. Sargent Florence in *Biological Aspects of Social Problems*, Ed. J. E. Meade and A. S. Parkes, Oliver & Boyd, 1965.

cheapness and simplicity may well be the answer for the poor, over-populated countries. Yet the developing countries India, Pakistan, South Korea, Tunisia and Turkey are the only ones with Governmental plans for assistance in birth control, and India with its 500 millions now is spending only one penny a head on the birth control campaign.

NEARER HOME

In Britain several bold and imaginative but limited, experiments have been made, to persuade 'problem parents' to limit their family size. In a survey sponsored by the Eugenics Society between 1948 and 1950 it was found that about half of these problem families had five or more children in them with consequent burdens upon public funds of one sort or another. Below is given one example of the work which has been done to improve the situation.

Dr Dorothy Morgan and a team from the Central Health Clinic, Southampton, started an experiment in 1961 to try to give definite help in birth control methods by home visits to a group of so-called 'problem parents', some of whom were mental defectives. Regular attendance at a Family Planning Clinic was beyond the capability of most of them. By 1964 116 out of the 150 parents concerned (76%) were using birth control methods. This is indeed a remarkable achievement because in the early days of the experiment resistance to any advice was high and strangers were suspect. In three years Dr Morgan noticed that some of the homes showed a definite improvement in their standard of living; wallpaper was replacing peeling distemper and chipped paint. The beginnings of pride in the home can be put down to the confidence of women in birth control to limit their family size and also to an increase in their time and energy spared from coping with a new baby.

Before the domiciliary service began 142 babies were born to the couples between June 1959 and June 1961; after the service began 32 babies were born to the same couples between June 1961 and June 1963, 110 fewer than in the previous period. The estimated saving to public funds on the 110 unborn children was £5,874. This sum is made up from savings on the maternity services, child allowances and cost of children in foster-homes, etc. Probably more important than this 'hard' data is the uncosted increase in human happiness and release of energy to look after the rest of the family properly and enjoy other things in life.

22: Destruction of Habitat

INEVITABLY in all countries population growth is bound up with the production of food and raw materials and with the growth of cities—all potential threats to man's wild habitat.

Great Britain is very highly urbanized and has the fourth highest population density in the world, an average of 575 people per square mile. Between 80 and 90% of its population is concentrated in towns and cities and only about 4% earns its living from the land. This handful of people produces enough food for about half Britain's population, the rest comes from abroad in exchange for manufactured goods. Cities are the money-spinning centres of the economy and these hubs of wealth—London, New York, Tokyo—are vital because large populations cannot be supported efficiently by subsistence farming alone. Curiously, in order to improve the economy of India, it will be necessary to drain 80% of the population from the land and concentrate it in cities.

In Britain every year as cities expand land is destroyed. In fact each year since 1951, thirty-six and a half thousand acres of farmland have been mopped up by urban development, and by 1970 land amounting to a fair-sized English county will have become urbanized. Yet surprisingly, as any train journey reveals, Britain is still predominantly rural, and the following table gives some precision to this impression.

TABLE 23

The major land uses of Great Britain in 1950 (after Best)

	Area (1,000 acres)	*Percentage*
Agriculture	45,240	80·5
Woodland	3,700	6·6
Urban development	4,070	7·2
Unaccounted for	3,190	5·7
Total	56,200	100·0

In developing countries space is plentiful but food is scarce and the population is growing fast (with no prospect of rich new-found lands to be conquered and developed). The economy is inefficient as has already been pointed out. In Africa, for example, the soil is poor and the climate unsuitable for rearing good grazing animals as in England. Nevertheless, game is being shot off to make room for domestic animals. Such mismanagement leads, as we shall see, to the decay of the soil and, if history is a guide, could lead in turn to the decay of populations.

While the primary threat to man's habitat is his unchecked fertility, the secondary threat is haphazard planning of his environment. The prime need here is to reconcile the conflicting needs of living space, industrial and agricultural development, the development of communications and the protection of areas where solitude and the beauty of colour and form can be enjoyed. To reconcile the vast artificial environment of man with the natural environment calls for long term ecological planning. Ecological because the planning needs to arise from an understanding of what makes nature tick and why; what happens for example to natural processes when they are exploited, interfered with or handicapped by man's activities. Only when answers to these questions have been obtained can planning of land-use on a local and world basis be soundly based. A few examples of this need for *ecological* foresight, based on planning and research are given in Table 24. They apply to the British countryside but the thinking in them could apply to many other parts of the world.

TABLE 24

*Human impacts on the countryside**

Activity or operation	*Area affected*	*Nature of effects arising*	*Incidence*	*Examples of problems and possible solutions*
Cultivation:				
Cereals Other field crops	Most arable areas	Ever increasing intensification means trend towards more rigid monoculture for each crop, and this tends towards elimination of some other forms of life. Use of certain chlorinated hydrocarbon insecticides on cereal seeds has been shown to have serious effects on birds. Some of these insecticides may also be having undesirable effects on soil and aquatic fauna and related food-chain organisms	General in arable areas, increasing	Areas of most intensive farming. Research at Rothamsted and Monks Wood Experimental Stations
Rubbish dumps	General near centres of population	Smaller communities cause most damage to amenity and wild life, by using wet-lands, quarries, sea coves and ravines. Dumps provide extra food and may thus be responsible not only for attracting but for making possible increase in population of rats, crows, rooks,	Serious, massive and universal problem, increasing rapidly in all populated areas	Swanscombe; Cors Goch;—keep Police and L.A. informed of abuse or irregularity. Larger operations (Reeth R.D.C., N. Yorks) L.A. projects subject to planning permission and appeal to Minister: Sheffield Corporation and Peak N.P. Board. Clearance of smaller un-

		jackdaws, sparrows, gulls. Problems increasing where disposal facilities lacking. Effluents pollute streams, ponds and ground water. Special problem of bulky and heavy refuse, e.g. old vehicles and other large metallic consumer goods, disposal of which is difficult through L.A.s		authorised dumps by C.N.T.s and by other volunteer parties, e.g. C.F.N. Conservation Corps
Salmon netting	Estuaries and neighbouring coastal waters	Destruction of and demands for elimination of seals	Limited to tidal waters and estuaries of certain rivers; season February–October	River Tweed; research on predator populations to determine criteria for control, e.g. marking and culling of seals at Farne Islands and Orkneys; Consultative Committee on Grey Seals and Fisheries
Spoil disposal, bings, tips etc.	Coal fields	Cover up all natural features and potentialities of land; destroy scientific interest of marshes, ponds or lakes; beach pollution, damage to coastline, amenity, wildlife; geological features often buried when quarries used for spoil disposal. Air pollution from burning tips	Widespread and serious in most coal fields	Bilston Glen; Fife Coast; Stodmarsh: rehabilitation by covering with soil and planting; bulldozing for industrial housing or recreational development

* Adapted from *The Countryside in 1970*, H.M.S.O., 1964.

RIVER POLLUTION

Man has always ruined and destroyed his habitat in his efforts to support his increasing population. The gradual poisoning of rivers since the beginning of the industrial revolution is a case in point. The black dyes of Leeds turned its river into a dark, filthy sewer. In 1862 Hugh Miller wrote about the River Irwell in Manchester:

> Nothing seems more characteristic of the great manufacturing city, although disagreeably so, than the river Irwell, which runs through the place . . . The hapless river—pretty enough stream a few miles up, with trees overhanging its banks and fringes of green sedge set thick along its edges—loses caste as it gets among the mills and print works. There are myriads of dirty things given it to wash, and whole wagon loads of poisons from dye houses and bleach-yards thrown into it to carry away; steam boilers discharge into it their seething contents and drains and sewers their fetid impurities till at length it rolls on—here between dingy walls, there under precipices of red sand-stone—considerably less a river than a flood of liquid manure.

Now, most rivers, lakes and tidal waters close to industrial areas are polluted by hot water, chemicals or organic wastes which use up oxygen, destroying the aquatic animals and plants. Chemical spraying of D.D.T. and similar insecticides has taken its toll of fish and their food. The Thames, for example, like so many British rivers, once abundant with salmon and oysters, has been turned into a desert where the dominant creature is a small red worm, *Tubifex*, which thrives in dirty oxygen-poor mud. And in the Trent in 1887, 3000 salmon were caught; after World War II, none. Not only is there a loss of fish but other creatures are affected too. The larval stages of caddis flies, stone flies and may flies, the food of young salmon, are destroyed quickly by D.D.T. and recover their numbers very slowly. Rivers are involved also in the wider problem of water demand. Both industry and domestic consumers with their ever-increasing requirements are creating a demand for larger reservoirs which take bottom land from agriculture, drown villages and perhaps destroy areas of natural beauty. And the rivers affect the sea. In the past century, for example, the Durham coast according to Dr David Bellamy has lost 90% of its seaweeds. Quantity and diversity of species have been affected. 'Diversity of sea-weed means a diversity of marine life for which it acts as food and cover. Cut one and you cut the other. This means less fish for the inshore fishermen to catch.'

HEDGEROWS AND PONDS

Not surprisingly, as a predominantly agricultural and urbanized country, Great Britain has the lowest proportion of forest of any of the European countries, so that most of our birds (once dependent upon forest habitats) and insects have come to depend upon the 600,000 miles of country hedgerows. Anyone from abroad who sees late Spring and Autumn hedges for the first time cannot fail to be impressed by their scent, colour and activity; the new green of the hawthorns in Spring, the fresh-minted dandelions on the verges, the lacework of cow parsley about which the orange tip ricochets, and in autumn, spindle, blackthorn, buckthorn and guelder rose embellishing the hedgerows with their fruits. These hedgerows give shelter to forty of the fifty birds species commonly found on farmland; yet the hedge is being grubbed up on a vast scale, especially in East Anglia and the east Midlands. Indeed, there is some evidence to suggest that between one-tenth and one-fifth of all British hedges have been destroyed in the last twenty years.

Not only have hedges been lost but ponds also—those magic places of childhood shadowed by willows and reeds hiding the coot and moorhen. In one mixed farm in Wiltshire of 500 acres, ten out of the thirteen ponds have been lost in the past eighty-two years and this figure is probably not exceptional.

This unconscious destruction of wild habitat has been reinforced recently by mass use of herbicides and insecticides. Pest control is naturally essential in the quest for human food supplies, health and comfort, but its use needs to be founded on a knowledge of ecology. Misuse of chemicals can throw out of gear the fine machinery of natural processes as we shall see. Herbicides such as 2.4–D are used to control weeds in cereal and other crops so that cornfields are no longer spattered with the colour of poppies, corn-cockles, cornflowers and corn marigolds. These bright but troublesome weeds have withdrawn to the sheltering hedgrows before the farmers' sprays and here they should be preserved. Admittedly their banishment to the hedgerows as bystanders has been the farmers' gain, but the natural habitat of the weeds has been tampered with and the consequences of this must be far-reaching; insects that fed on the weeds are deprived of food and this naturally affects other links in the food chain, especially in the caterpillar feeders such as cuckoos and various song birds.

POISONED EGGS

Besides herbicides there are two main groups of insecticide in use: the organo-phosphorus and the organic chlorides dieldrine and D.D.T. The first are not usually persistent, but the others are and are therefore more dangerous. For example B.H.C., an organic chlorinated insecticide, persists for 14 years or more and tell-tale traces of organic chlorine compounds have been found in the eggs and flesh of sea birds like oyster-catchers, shelduck, kittiwake, puffins and others. This has serious consequences because such poisoned eggs are not fertile and thus bird numbers are cut down. The eggs of moorland birds like grouse, raven and golden eagle tell a similar story. Traces have been found too in the muscle of fresh water feeders like the dipper, heron and coot (see Fig. 41). All this evidence points to the fact that Great Britain is contaminated with chlorinated compounds, and that they are present in the food chains of these birds. As they go through the food chains the poisons concentrate. Some evidence for this statement comes from fish-eating birds of lakes. These birds contain most of the poison, the plankton-eating fish rather less, and the plankton much less. The upshot of this story of poisoning which reads almost like a Grimm's fairy tale with no happy ending is that the population of songbirds is declining because it is being poisoned or sterilized by eating contaminated food. Colour is vanishing from hedgerows and verges because sprays are killing off some of the loveliest plants, and butterflies like the blues of the chalk Downs are now quite scarce because sprays and ploughing are destroying their habitats. Grass feeders of the wayside like the meadow brown, marbled white and the small heath have been hard-hit too, not only by spray but by the total impact of modern agricultural methods on the countryside. Fig. 42, prepared by Dr Norman Moore, summarizes some recent parallel trends in British agriculture which affect animal habitats.

What do these facts mean for the future of Britain's wild life? They strongly reinforce what was said earlier about the basic need for ecological planning and basic research. Without these there cannot be good land use either in countries which are short of space like our own or in developing countries like Africa.

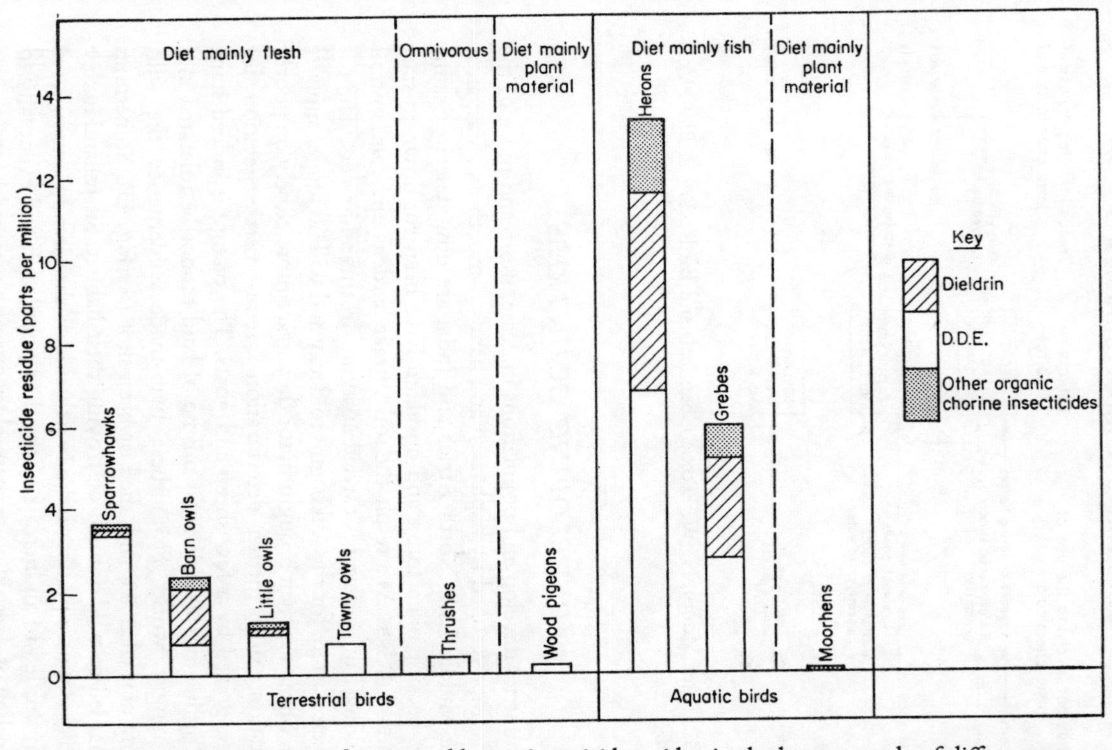

FIG. 41 Average concentration of organic chlorine insecticide residue in the breast muscle of different types of birds. The following were investigated (number of specimens analysed given in brackets): sparrowhawk (5), barn owl (9), little owl (7), tawny owl (5), thrush (4), wood pigeon (6), heron (7), great crested grebe (4), moorhen (6).

After Moore, N. W., and Walker, C. H., 1964, 'Organic Chlorine insecticide residues in wild birds', *Nature*, March 14th, 1964.

SOME RECENT PARALLEL TRENDS IN BRITISH AGRICULTURE WHICH AFFECT ANIMAL HABITATS

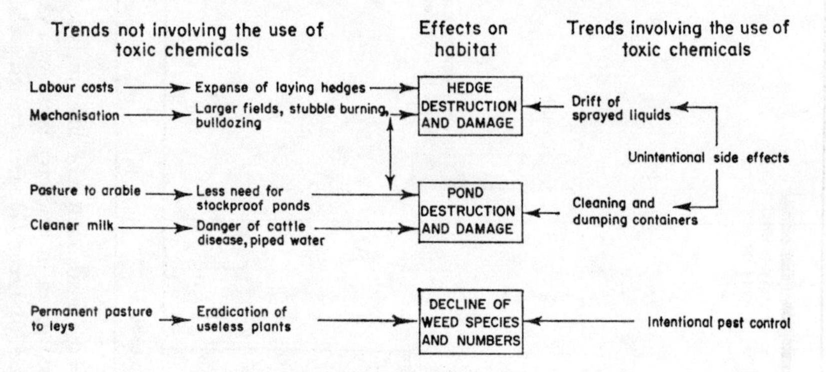

FIG. 42 After Moore, N. W. 'Toxic Chemicals and Birds', *British Birds*, Vol. 55, 1962.

DEVELOPING COUNTRIES

As countries develop and populations increase, more and more of the land is impoverished by haphazard and hand-to-mouth use. Vast areas in the Middle East, North Africa and India are now desert or semi-desert because open forest and bushland have been cut down to start up cattle ranches. As a result the soil surface hardens and rain, instead of sinking, runs off and is carried away in streams and rivers. The rot has then started, for the soil begins to decay and will no longer support cattle. Sheep are then brought in for they graze more closely on poorer vegetation. With further deterioration a semi-desert vegetation develops, grazed by a few sheep and goats. The picture is indeed stark and frightening for this gross misuse of land cannot be tolerated in a land-hungry world. From these man-made wildernesses the wild grazing animals have fled, died or been shot (see Fig. 43). Sometimes this has led to wild animals destroying their habitat. Sir Julian Huxley cites a case where pumas on the North side of the Grand Canyon were killed; this led to an increase of the deer population and eventually to the decline of deer who ate themselves out of house and home almost to the point of extinction. What needs to be done? More and more ecologists want to preserve whole natural communities of plants and animals, not as living museums for people of the year 2000 to gape at but as sources of wild protein to feed hungry people. An important question for ecologists to answer is why these natural communities produce magnificent wild animals while domestic animals kept by natives on the same land are often scraggy and in poor shape. Sir Julian Huxley writes:

> In eastern Africa the assemblage of splendid large mammals and birds, the last remnant of the climax community of pre-human evolution, is one of the world's unique enjoyment resources. But it is of immediate financial value, through tourism, to the local inhabitants. It is also of physiological value; large areas of the dry savannah lands of the region degenerate if cultivated or used for grazing cattle. But if they are properly managed, their communities of wild animals yield large amounts of 'wild protein' for human food—larger than can be obtained from domestic stock.

Probably one of the reasons for the good state of wild animals compared with domestic animals is that cows, for example, feed on certain kinds of grasses only and cannot use much of the available food. Indeed, there is not enough of it and it is quickly destroyed. The wild animals feed selectively on different plants so that the total animal population uses all the food resources of the environment but does not destroy them. Once farmers realise that *profit* can be made by ranching game animals as well as cattle (as is happening in Rhodesia and South Africa) then the people of the developing countries will be better fed and the vanishing animals of Africa (and other countries) will be preserved.

Destruction of habitat has taken place throughout history. The Phoenicians cut down their cedars to sell them to Egypt and to build their own fleets; the Minoans turned their forests into ships and ruled the Mediterranean. Overgrazing over thousands of years by sheep and goats destroyed much of the vegetation (once forest) of Syria and Palestine, Greece, Italy and North Africa. Tunisia, once a granary of Rome, is now semi-desert. And as the soil decayed and the habitat was destroyed so society decayed. For it was always the leaders who departed to richer places. Now there are no new-found lands to depart to. We have to learn to live with the land we have with a continually expanding population and this means world planning not only to give basic calories, vitamin fuels and minerals but to preserve beauty and peace.

WORLD DEVASTATION OF

FIG. 43 Places where the worst devastation of wildlife has taken place, showing some of

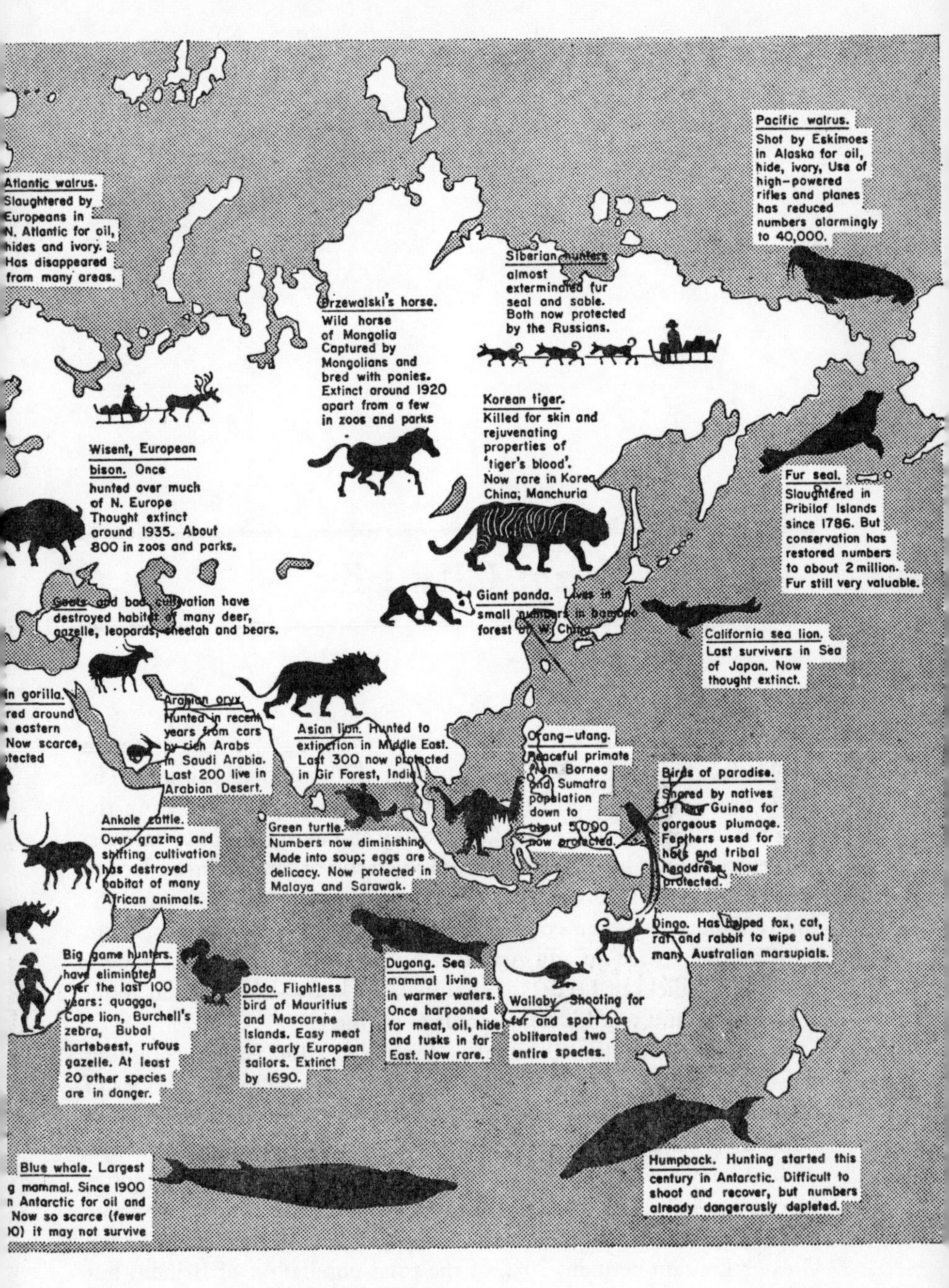

WILD LIFE
the animals in danger and some already extinct. After *Observer Magazine*, 17 April 1966.

23: City Life

THE growth of cities is perhaps the biggest single problem facing man in the second half of the twentieth century. The drift to the cities has gone on over 5,000 years and from the original fusing of shrine, citadel, workshop and market the city has been the centre from which men have been ruled and organized into effective co-operation. Thousands of years ago in Mesopotamia, Egypt and the Indus Valley, city men co-operated to control flood, to repair storm damage and to set up and maintain a vast irrigation system. Nearer home in Sheffield or Leeds, for example, the know-how of men about steelmaking and textiles was organized by the city.

The pull of city life has intensified in the last 150 years throughout the world; from 1800 to 1950 the number of people living in cities of 100,000 has increased twenty times from 16 millions to nearly 314 millions. It has been estimated that by the year 2000 only a tiny fraction of a country's people will be needed to produce the food it requires (about 8 to 12%) and the rest will work in money-spinning cities whose long-range effects on the countryside could be devastating. Today Greater London contains 20% of the total population of Britain; but this is as nothing compared with a city like Calcutta which in A.D.

2000 might contain as many people as there now are in Britain—about 60 millions! Already Calcutta is breaking down with its present six millions: Tokyo, Peking, Berlin, Paris, London, New York, Chicago, Rio de Janeiro, each with more than three million people, will become sprawling monsters if the trend goes on (see Fig. 44).

The social and biological implications of such 'man heaps and machine warrens' must concern us here. In bare bread and butter terms, for every thousand newcomers to a city the following are some of the extra facilities required:

an additional 36·5 million gallons of water each year;
additional sewage and treatment facilities for 62,050 organic water pollutants a year, or 300 new septic tank leaching systems;
about £25,000 extra for dealing with air pollution; one extra hospital bed;
1,000 new library books;
4·8 primary school rooms, 3·6 secondary, and 8·8 acres for playing fields;
1·8 more policemen, 1·5 more firemen and a fraction of a jail cell.*

For Calcutta at present, and for the majority of cities in developing countries these statistics are day dreams, as anyone who has witnessed the plight of Calcutta will agree. Here millions of people sleep in slum shacks or on streets and millions more fight the running battle between starvation, disease and death. The plight of India is so great that a recent W.H.O. report stated that up to £7,000 million would be required over a period of 25 years, ending in 1975, to house the new inhabitants of cities with over 100,000 people. And the provision of city-wide services, utilities and transport would at least double the figure.

The proximity of such densely populated cities such as Calcutta to the open countryside is a serious worry to those concerned with world health. Such cities contain millions living in insanitary, filthy conditions. They are places where plague as rampant as the black death could return. Man living under such insanitary conditions might come into contact with the wild rat and get bitten by plague-carrying fleas or the urban rat might mix with rats from the countryside and transmit the plague inside cities. W.H.O. officials recently voiced their alarm as spreading cities in Asia and Africa bring the city rat closer to wild rodents. It is true that modern medicine and hygiene will control the disease but only 20 years ago around 57,000 Indians died of plague in one State

* American data.

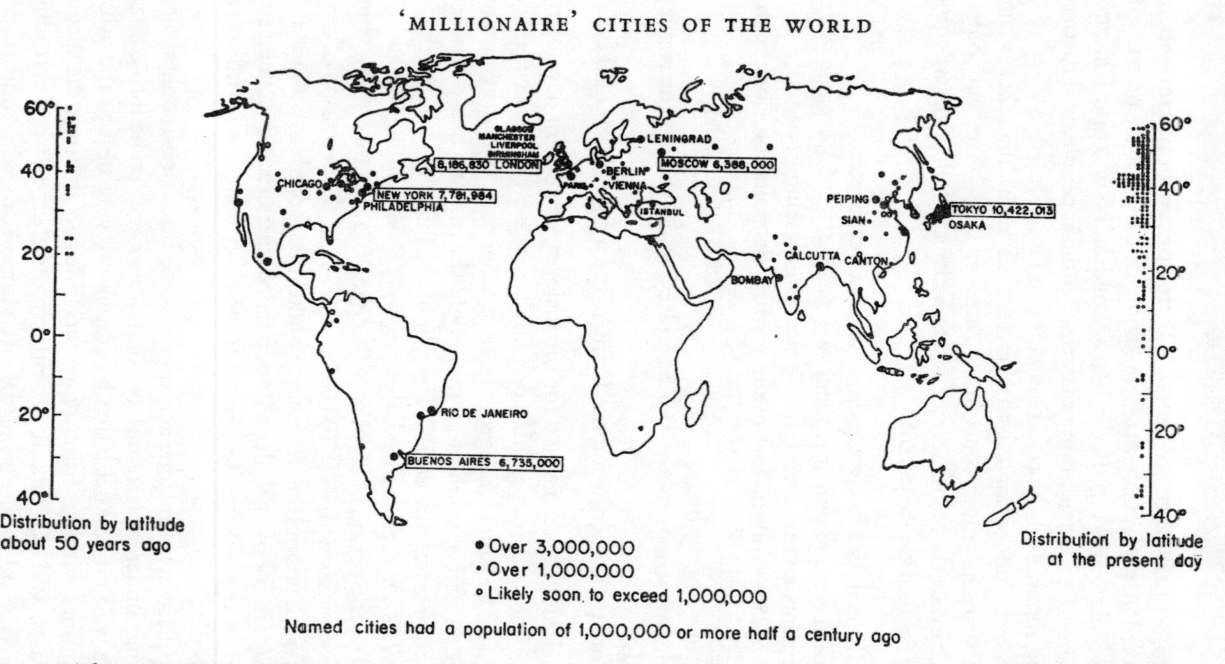

FIG. 44 This map illustrates the increase in the number of large cities (over one million inhabitants) in the last fifty years and their distribution by latitude.

After *Man and his Environment* (B.B.C. Publications, 1965).

alone. Now spreading urbanization has created the conditions for an explosion of disease.

The repercussions of city growth, hinted at above, are felt far into the countryside. It is not only the raw and spreading edge of the suburbs, the pylons, the filling stations and transport cafes but the mundane yet vital need for water, sewage disposal, aerodromes and trunk roads, and how to cope with air pollution. Only four years ago in Great Britain in 1963, over a million tons of smoke and 100,000 tons of grit were fouling the air from private grates and wafting over the countryside to blacken trees and prevent them respiring. Reservoirs, aerodromes and huge roads sterilize great quantities of good land. Last but not least, there is the problem of traffic. The Buchanan report tells us that in 1962 there were 16·4 million families in Great Britain, 6·6 million cars and 1·8 million motor bicycles. Not many families in fact owned a motor vehicle but many incomes at that time were verging on the car-affording class. The report predicted:

in 1970, 12 million cars, 18 million vehicles (lorries, buses, motor cycles and cars);
in 1980, 19 million cars, 27 million vehicles;
in 2010, 30 million cars, 40 million vehicles.

The greatest pressure on our roads and in our towns and cities will come in the years up to about the mid 1970's when half the rise predicted by Buchanan will be upon us. The motor vehicle will be with us for a long time; it is the towns that will have to adapt themselves to the traffic increase. Unless positive policies for adaptation are put forward and implemented by the mid-1970's we should have 15 million cars on the roads with all the resulting confusion, the wasted time and money (see Fig. 45), the injury and death by accident and the ugliness. For the motor-car could become a kind of multiplying parasite which so breaks down standards that ugliness becomes tolerated, while our whole historical and architectural heritage falls into neglect and disrespect. The first crack in our will to resist this plague could come with permanent parking of vehicles at the kerbside. Then the oil stains, battered kerbs, garbage, broken rails, maintenance of cars on the street would become accepted, until ugliness would be the normal and beauty the curious exception to the rule.

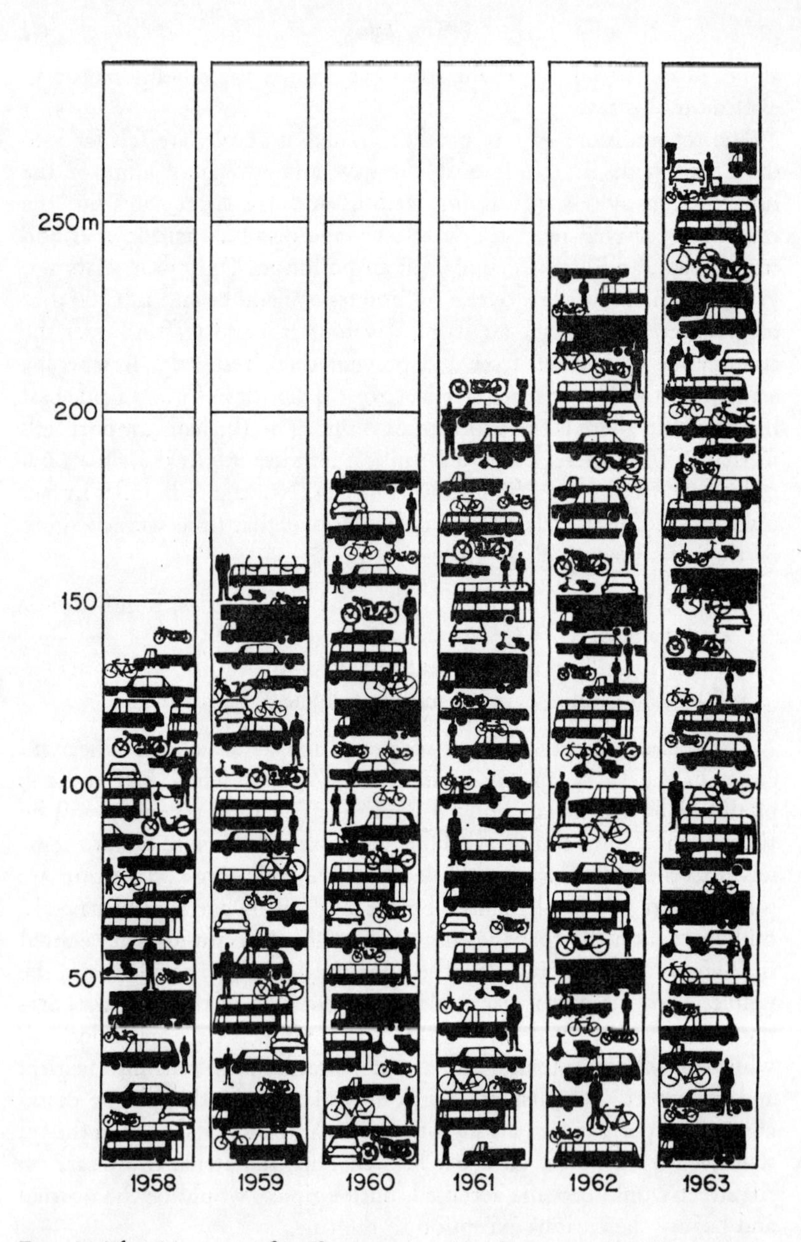

FIG. 45 The rising cost of traffic congestion in urban areas. (Based on calculations by D. S. Reynolds and S. G. Wardrop of the Road Research Laboratory.) After *Traffic in Towns* (H.M.S.O., 1963).

LOOKING FORWARD

The huge cities that we see today—Manchester, Leeds, Birmingham, were generated by the mine, factory and railway of the nineteenth century. Towns and cities formed rapidly and clustered thickly on the magnet of the coalfields where the new industries, iron and coal mining, smelting, cutlery, hardwear and glass manufacture, started to flourish. During the first half of the nineteenth century Manchester grew from a town of 85,000 to a city of 400,000 at the hub of the Lancashire cotton industry. Leeds grew from 53,000 to 172,000 and Bradford from 13,000 to 105,000 as a result of the mechanization of the cloth-weaving industry.

The railway line, too, pulled people to it so that there was a huge population build-up in the manufacturing towns and cities along its track. In the age of steam big towns and cities were inevitable, but now in an age of jet travel and rapid communication generally, when distance is not of great importance, huge cities are like dinosaurs living on in a fast-moving epoch. Smaller communities of for example, 50,000 people, can have the industries and the goods, the libraries and the concerts and all the sense of being part of a larger concourse, which only giant cities had in the past. These smaller units would avoid all the problems of traffic jams and the wear and tear of commuting with its frustrations and planners could see to it that the individual was not cut off from nature or dwarfed by huge buildings, as are millions of people today. As the Buchanan report says there is a need to create or recreate towns which in the broadest sense are worth living in—'convenient, variety of choice, contrast, architecture, history visible in the buildings' are all needed.

THE DANGER OF CROWDING

In the past it was important for the city dweller to be able to live and breed under conditions of crowding. It is possible that established city dwellers have, over hundreds of generations adapted physically and mentally, that is genetically, to these conditions. But the revolution in health, as we have seen, has made adaptation to the old killing diseases redundant at least in rich countries. It is the illness brought on by the state of mind which is likely to increase in big cities nowadays. Gigantic buildings dwarfing the human scale, lack of peace from the hustle and bustle of people and traffic; the complexity of new machines which need continuous fixed attention, unrelieved by physical activity, loneliness and the inability to get on with very unlike people, can all bring about the tension states, neuroses and psychosomatic illnesses we hear so much about today. We should not forget that man is primarily adapted to living in small well-knit groups as in village life. In cities, even now, these ingrained, adaptive traditions remain in the small well-defined neighbourhoods or streets of a town. When these small, face-to-face communities are broken up, as sometimes happens when slum dwellers are moved to housing estates (see Chapter 14), people are liable to become unhappy and ill.

As cities increase even further in size, if observations and experiments on animals are anything to go by, the stress arising from overcrowding could lead to mental instability, suicide (see Chapter 27), and perhaps social and political disturbances. In rabbits and many other mammalian species populations build up and then die off or 'crash'. Examination of the dead bodies shows no evidence of disease or starvation. It is possible that they may have been killed by over-activity of the pituitary and adrenal glands brought on by stress. Abnormal social and sexual behaviour has been shown in experiments with rats which were kept at twice the density compatible with a healthy population. Females became progressively less adapted to nest building and eventually stopped building them altogether. Some of the males became homosexual, others passive zombies, never fighting or showing sexual interest, and because they were so meek and mild they had to rise early to be able to eat and drink in peace!

Whether the manifestations of 'pathological togetherness' will ever affect humans in this way only time will tell, but some evidence that it may do so comes from the work of an American psychiatrist, Professor Rene Spitz. He found that mothers forced to live in crowded conditions are often unable to make and maintain proper emotional bonds with their babies. And this failure in the essential tie between mother and child damages the normal emotional development of the child.

Another result of over-crowding is that social order may break down into a dictatorship. In animals this results in a 'peck order' in which No. 1, the dominant hen, pecks the rest, No. 2 pecks all but No. 1, but is herself pecked by No. 1—and so on. The establishment and maintenance of a peck order is a stressing business and the greater the crowding the worse it is, for disturbances in the hierachy are constantly being set up by competition. This peck order is seen in any human organization; when the boss takes it out of one of his underlings, the latter inevitably directs his attack to weaker individuals, his wife or children! Crowding would almost certainly aggravate the pecking instinct in man. In normal children too, fighting, snatching, and breaking toys increases under crowded conditions while brain-damaged children show extreme aggressive behaviour such as biting. All of us have felt aggressive in crowds. All of us living in cities are often fatigued with the surfeit of social relationships. Most of us resent a tap on the door or a ringing telephone in the evening. Whether animals have such feelings is, of course, unknown, but voles in high-density populations have many more battle scars and are in poorer health than animals living in less dense populations.

If crowding ever forced an ant-like dictator state upon people in cities, that is human beings condemned and restricted to perform endless and similar operations without thought hour after hour and day after day, then human talent which thrives on novelty, would be on the scrapheap. Such conditions might even reduce the chances of man for a reasonably long existence on this earth.

It is clear that Megalopolis, the city of millions, is self-defeating with its utter confusion in communication, its special health hazards, its sheer size which allows only the richest to get far away from crowds and noise. Each one of us now belongs not to a particular bit of countryside or town or city, but to a social organization which needs to use a whole region as its living space—villages, forests, water, towns. And each region needs planning by biologists, sociologists, geographers and architects if land is to be used properly and with foresight. We must never forget, the humble truth, that since the beginning of the neolithic period man has lived in small communities and most ordinary people still wish to do so and are psychologically adapted for doing so.

Part 5: Youth and Age

Gangs, crime, illegitimacy, V.D. and drug taking are some of the violent marks of youth on society. At the tail-end of life lies the urgent problem of medical and social care of the old. Nowadays most people are living for three score years and ten, and are likely in the future, to live longer, perhaps in a semi-artificial state. Youth demands attention by the very vigour of its biology, but the relatively silent problem of the old and lonely is apt to be overlooked. The chapters on youth spotlight some of the more depressing features of the behaviour of a minority of young people, but it is obvious to those who know the young that the majority are capable of enormous idealism and thoughtfulness. They will march in support of nuclear disarmament and help old people with gardening and domestic chores. Down the ages alarmists have claimed that the behaviour of the young was wanting or even outrageous, but the unknown person who said 'society, sir, has been so long going to the devil that it is devilishly odd it has not yet come there' has much to recommend his sense of perspective. Perhaps too, the behaviour of the under-twenties is influenced by the creations of the over-forties. After all the young have 'never known a world without violence on television, or carnage on the roads, or money to burn in one's pocket, or sexual license in public places'. These are not their creations but those of their elders.*

In writing this section the sociologist's territory is invaded, but sociology and human biology have no clear boundaries; a man's hereditary endowment can determine his behaviour and human behaviour makes society what it is. Very recent research, for example, has shown that sex chromosome abnormality is apparently linked with some types of violent and anti-social behaviour. Does this mean that after one act of violence courts of law should hear chromosome as well as psychiatric evidence?

The unique nature of the 1,300 Chicago gangs of the1920s's or of our own Mods and Rockers was and is, due to novel collections of personalities drawn to particular environmental circumstances, particular combinations of genes fitting particular niches in society. Genetics cytology and human biology may help to explain these and other aspects of human behaviour to give a foundation to the science of society.

* Eldon Griffiths, M.P.

24: Crime

THE PROBLEM

CRIME in Britain (other than non-indictable motoring offences) is a characteristic of young males, more particularly the 14 to 20 year old. In 1955 for example about half of those found guilty of housebreaking and a third of those caught thieving were under 17. The phenomenon of the young criminal is as old as the hills but since about 1955 there has been a marked increase in the amount of house-breaking and violence with accompanying gang-life and group hooliganism, especially among boys in their last year at school and early working life. Particularly striking is the great increase in crimes of violence—from 1,583 convictions in 1938 to 11,519 in 1961. Of these convictions, youths under 21 formed 17·6% in 1938 and 41% in 1961. The same story is true in the United States (where the number of arrests of people under 18 was 9% greater in 1962 than in 1961) and of the western world generally.

It might be argued that growth in crime and population are running parallel and that there is no real growth in crime. The facts belie this. For every 100,000 of the British population in 1900 there were about 250 crimes; in 1965 there were 2374.

Thirty years ago unemployment, overcrowding and poverty seemed to be plausible explanations of anti-social behaviour, but since the war crime has grown against a background of plenty with full employment, good wages, new houses and schools.

Twelve possible factors (which could interact) which appeared to relate somehow to criminal behaviour were examined critically by Barbara Wootton in 1959. The factors were:

1. The size of the delinquent's family
2. The presence of other criminals in the family
3. Club membership
4. Church attendance
5. Employment record
6. Social status
7. Poverty
8. Mother's employment outside the home
9. School truancy
10. Broken home
11. Health
12. Educational attainment.

It seemed that most law-breakers came from large families and not infrequently other members of the offender's family had been in trouble with the police. It was more than likely that an offender did not go to church but he might or might not belong to a youth club. He was often a poor worker in a dead-end job in an uninspiring district and came from the lower social classes and a broken home. No evidence was found that his crime was connected with his mother going out to work. His health was not generally poor; often he had been in trouble at school and played truant there. To this patchwork from which violence emerges could be added the fact that leisure time is increasing and also money with which a youth can buy independence from his parents. After paying his bed and board he is free to spend the rest of the money as he likes and the barrage of advertising invites him to waste it. He has no responsibilities for raising a family or buying a house and his leisure time is often wasted simply because he does not know how to use it constructively. His only desire it seems is to relieve his boredom by excitement and thrills, even the thrill of breaking the law.

CRIME AND DESTINY

Can the personality of the delinquent help to explain his behaviour? Is there a flaw in his personality which given the appropriate environment (a broken home, unsuitable companions) expresses itself? Put another way, given the right set of circumstances does his genetic endowment determine certain physical and emotional characteristics which predispose him to crime? As far as personality is concerned, many offenders seemed bored with life, inarticulate, impulsive, unreliable, with poor judgment and no conscience. A number of these characteristics and other emotional and intellectual qualities which determine behaviour have been shown to be genetically determined.

The study of the heredity of crime stems from the work of Lange already mentioned in Chapter 5. He examined 30 pairs of twins (see Table 25) for their criminal records and in general his work supports strongly the idea that the genetic component in crime is high.

TABLE 25

Record of Lange's Twins

With crime record	*Type of twin (all of same sex)*		
	One-egg	*Two-egg*	*Doubtful*
Both	10	1	1
Only one	2	15	1
Total	12	16	2

From C. D. Darlington, *Genetics and Man*, Allen & Unwin, 1964.

Identical twins are genetically identical since they developed from the splitting of a single fertilized egg. Fraternal twins develop from two separately fertilized eggs and are no more alike genetically than brother and sister. If identical twins are brought up separately, as sometimes happens when they are orphaned, they are likely to be exposed to different environments and these different environments will act on a constant hereditary endowment. The relative importance of the environment and the genetic outfit can therefore be determined in respect of physique, physiology and more important, social character and behaviour. Lange found that 'identical' twins had criminal records that agreed closely; not only was the type of crime similar, but the age at the first offence was also very close. If one twin was an embezzler, his twin might be a forger, but not a burglar. One pair studied had no head for drink—it made them fighting mad and they tried to knife people. Another pair 'were crooks of genius, quite unusual swindlers'. Fraternal twins (not identical genetically) were no more alike than ordinary brothers and sisters in respect of crime. The one pair where both took up crime, shared widely different criminal records in quantity and quality. One twin after two lapses that sent him to prison had been a respectable person for years; the other had been a common vagrant and professional thief for 20 years. Two of Lange's pairs of identical twins had different records. How is this explained? Rather than confusing the notion that a predisposition to a criminal way of life is inherited it strengthens it, for just as a criminal can inherit a faulty nervous system that predisposes him to crime, violent injury to the head (and therefore to the nervous system) can bring about the same behaviour. This is what happened in the case of Lange's two pairs of twins, for their differences in behaviour were due in one case to a knock on the head and in the other a head injury at birth. Since Lange's time further studies have been made of hundreds of pairs of identical and fraternal twins. These have shown that if one of a pair of identical twins is a criminal, his co-twin has about double the chance of becoming a crook as he would have if the twins were of the fraternal type. As well as crime, behaviour disorders of childhood, homosexuality and, to a lesser extent, alcoholism are strongly inherited tendencies. The evidence for juvenile delinquency is not so impressive. Table 26 gives a summary of the evidence from twin studies for various types of criminal, anti-social and asocial behaviour.

It has been argued that the traits for criminality are strongly inherited, but they are obviously not clear-cut characters like O and A blood groups. Rather, it would seem, criminality is a continuous character like intelligence, fertility, height and weight, which is controlled by large numbers of genes having plus and minus effects on major characters. But what are these major characters? Obviously there cannot be genes for 'criminality'. It does seem, however, that aggressiveness, impulsiveness and brawn are inherited traits which given the right circumstances could lead to criminal activity.

That many genes controlling such traits are involved is suggested by

TABLE 26

Concordance of identical and fraternal twins respectively for various types of criminal, anti-social and asocial behaviour

Type of behaviour	*Number of twin pairs*	*Identical*	*Fraternal*	*Proportion concordant*	
				Identical	*Fraternal*
Adult crime	225	107	118	71	34
Juvenile delinquency	67	42	25	85	75
Childhood behaviour disorder	107	47	60	87	43
Homosexuality	63	37	26	100	12
Alcoholism	82	26	56	65	30

From H. J. Eysenck, *Crime and Personality*, Routledge, 1964.

the fact that criminals vary among themselves, from those who go to prison once and never again to the old lag who is always in and out of prison. Moreover, many offenders come from families who are 'known' to the police. These variations presumably are due to different proportions of plus and minus genes distributing the traits at random as the chromosomes are shuffled in gamete formation.

Very recently Dr C. E. Blank has produced some hard evidence which links anti-social behaviour with a sex chromosome abnormality. Dangerous male criminals in two mental hospitals showed a high proportion of individuals (300 per 10,000) with an *XYY* chromosome complement. The extra *Y* chromosome seems to cause aggressive or violent behaviour and disturbs normal patterns of growth (see Plates 3a and b and Chapter 3). Further an *XXYY* constitution is very rare in the population at large but occurs in about 60 or 70 of every 10,000 patients in mental hospitals like Rampton and Broadmoor. Even more alarming, the *XYY* abnormality has been discovered among men in an ordinary short-stay prison.

*Mainly against property rather than against persons.

What can be done to prevent crime if the 'determinants' do form an important link in the chain of events leading to crime? As yet we do not know sufficient about the releasers to criminal behaviour to knit them into a general theory of crime. But we do know by twin studies that a man of a certain genetic constitution, given the right environment, has a good chance of becoming a criminal, and we know, too, by the study of heredity that criminal habits are propagated and that every society will receive its share of law-breakers. It could be argued that certain individuals, regardless of their environment will continue to steal and be violent. Men with an *XYY* constitution are often first convicted in boyhood and thereafter many times reconvicted. Prison will protect present society from them and future society could be protected by preventing them from having children. Legalized abortion would have much the same eugenic effect as this. It is worth noting too that the worst criminals are willing to risk being shut away in prison for long periods. Does this mean that, inherently, they can do without women and do not want children?

Although a programme of eugenics, which suggests that man should use his scientific knowledge to prevent children being born who are likely to be severe social burdens, might lighten the load on society, such a programme would never be tolerated at present.

25: Gangs

IT was pointed out in the last chapter that both gang life and hooliganism appear to have increased in recent years. It is indeed commonplace nowadays to see groups of youths hanging about in an aimless way as if waiting for something to happen. Gangs of course are not new phenomena, nor are they found only in the slums of cities, nor are they necessarily bad. Every place has a gang—really a society that provides youth with an escape from the world of parents or insipid work conditions.

THE NATURAL HISTORY OF GANGS

Group behaviour—gregariousness—is an essential part of man's make-up just as are self-preservation, sex and getting enough to eat. From an evolutionary point of view a gang or group or herd enlarges the advantages of variation in the individual. There are some human beings whose individuality is not immediately favourable and departs widely from the norm and a group gives such variation chance to survive by the increased protective strength of the larger unit. In fact to compare the gang with a herd of animals is not too far-fetched. Gangs, especially in attack or in intimidation, act as 'a single creature whose power is greatly in excess of the sum of the powers of its individual members'. Fifty Rockers' motor bikes can make a formidable pack of steel and set up a roar big enough to frighten the timid. As for the 'lower' animals:

> The wolf pack forms an organism . . . stronger than the lion or tiger; capable of compensating for the loss of members; inexhaustible in pursuit, and therefore capable by sheer strength of hunting down without wile or artifice the fleetest animals; capable finally of consuming all the food it kills and thus possessing another considerable advantage over the large solitary carnivora in not tending uselessly to exhaust its food supply. The advantages of the social habit in carnivora is well evidenced by the survival of wolves in civilized countries even today.*

Gangs, whether they are of Mods, Rockers, or Teds, provide the thrill and zest of joining in common interests—hunting, fighting, capture, flight and escape. Fighting with other gangs and with the world in general gives boys the opportunity for most of these exciting group activities. A gang needs conflict too to weld it into a piece; it therefore thrives on warfare, planning for warfare and protecting its territory and possessions.

All gangs are unique in that they are made up of human beings each genetically novel and each gang has a definite structure, definite interests and activities. There is usually an inner circle made up of the leader and his henchman who create the right environment for the rest to obey them: these are the less enterprising and less capable rank and file and the hangers-on who hesitate to go the whole hog in group activities. The gang, like a herd of animals, has a structure imposed by the necessity of defending itself against other gangs and for efficiency in attack. But gangs are essentially a need of youth, for marriage ultimately disperses them except for a persistent hard core of criminal types; so the gang that once supplanted home, religion, work and play succumbs eventually to home and family.

Most of this was written 40 years ago about the 1,300 gangs of Chicago and it appears to be true of the gangs of today although the dress and trappings of the modern gang have changed from the cloth-cap and razor of the inter-war years to the leather jacket, motor-bike, cosh and flick-knife of today.

*W. Trotter (see Bibliography).

BODY BUILD OF A SAMPLE OF OXFORD STUDENTS, SANDHURST CADETS AND BORSTAL LADS

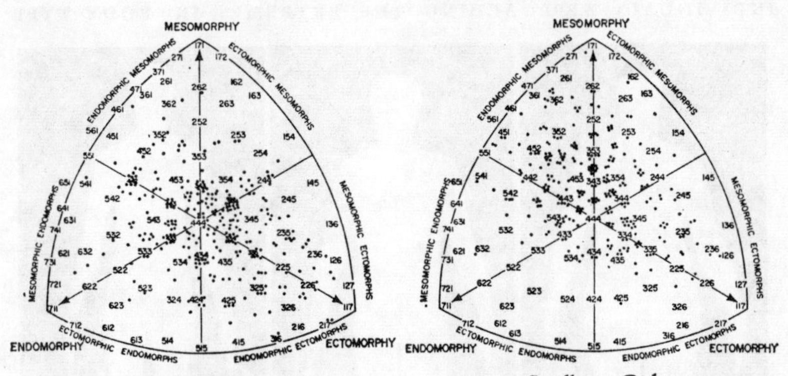

Oxford Students *Sandhurst Cadets*

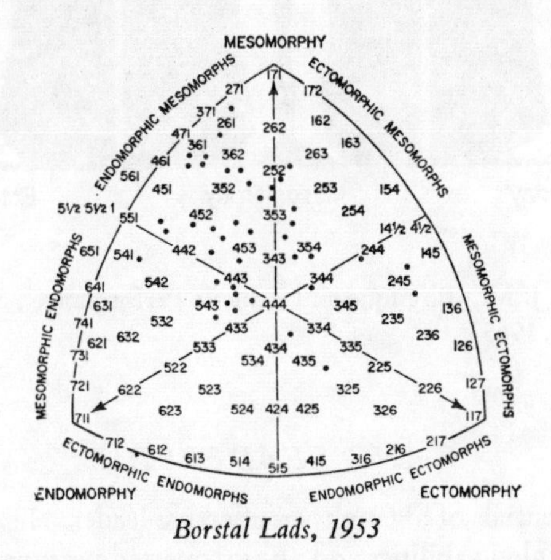

Borstal Lads, 1953

FIG. 46(*a*)

Oxford Students

Note the fairly even distribution of body types with perhaps a slight increase in the lower right-hand corner signifying bean-pole types.

After Tanner, J. M., *Human Biology* (O.U.P., 1964).

Sandhurst Cadets

Note the large number of brawny, muscular types.

After Tanner, J. M., *Human Biology* (O.U.P., 1964).

Borstal Lads

Note the large number of thick-set, brawny, athletic lads and the almost complete absence of the bean-pole type of youth.

After Gibbens, T. C. N., *Psychiatric Studies of Borstal Lads* (O.U.P., 1963).

INDIVIDUALS APPROACHING THE EXTREMES IN BODY-TYPE

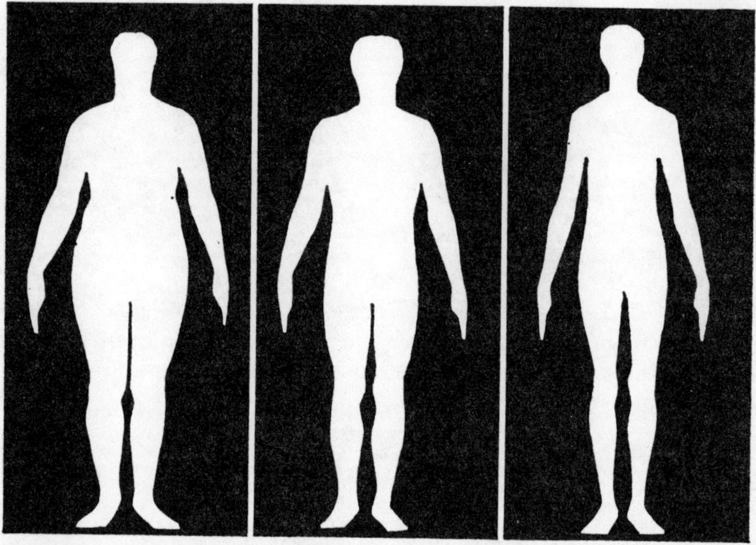

FIG. 46(*b*)

After Tanner, J. M., 'Physique and Athletic Performance', *Science Survey B.* (Penguin Books, 1965).

GANG LEADERSHIP

One of the essentials of a gang is an effective leader. He needs the right physical and athletic abilities and an extroverted personality—one who is assertive, dominant, impulsive, given to physical adventure and taking risks. The physique that goes with this type of personality is usually thick-set and brawny and allows him to behave in an extroverted way. Body-build is certainly determined by heredity and there is some evidence, again from twin studies, that an extroverted personality is inherited, and that build and personality are linked. The diagrams in Fig. 46a compare the body-builds of young men at Borstal, Oxford and the Royal Military Academy, Sandhurst, and it is clear there are more brawny lads at Borstal and Sandhurst than at Oxford where the beanpole type predominates slightly. Perhaps this is because careers in the Army and in crime have similar expectations—'power, risk, action, crisis—it is the object towards which the energy is directed which is different'.

While many gang leaders and gang members are tough, others of poor physique can lead groups. In Chicago it was found that the hunch-back or the dwarf could dominate healthy boys provided he had imagination and intelligence. He could often 'think up things for the other boys to do'. With a leader who can dominate a group and with each member fitting into a particular niche (doing a particular job) the gang in action can be a formidable unit.

Violence however is happily rare amongst youngsters although it is magnified by some newspapers, and some individuals who forget the disgraces often committed on Boat Race night by the 'upper classes'. One wonders whether youth has really changed much. It was noted in Chapter 14 that neurosis, supposedly a symptom of the affluent society, has probably not increased since the eighteenth century, and the next chapter presents some evidence, perhaps slight, that teenage morals have not changed much since Chaucer's time except that youth is now healthier and reaches maturity two or three years earlier than it did.

Man is instinctively susceptible to leadership and there is no doubt that the energy, vitality and craving of youth for adventure, danger and excitement needs to be profitably channelled in the right direction through experienced and sympathetic leadership.

26: The Spermatozoon and the Spirochaete

BESIDES the growth in crimes of violence and gang hooliganism among teenagers, there has been an increase in V.D., especially among young girls. The number of illegitimate babies born to women under 20 has also risen. Both these facts seem to point to an increase in precocious sexual behaviour among teenagers. What is the detailed evidence for this statement? In England and Wales in 1956, 887 illegitimate babies were born to girls aged between 11 and 16, and in 1961 the number had nearly trebled (2,534). Although the number of girls in this age group increased by only one-fifth. The total number of illegitimate births to mothers under 20 was 9,688 in 1960 and had risen to 13,938 in 1962.

The extent of promiscuous behaviour can be gauged by the facts on V.D. Promiscuity means 'indiscriminate mixture' As V.D. is sexually transmitted, the spread of it must involve at least three people: one or both partners in the sexual act must have had intercourse with someone else and therefore venereal diseases are linked with promiscuity. Teenagers between 15 and 19 in England and Wales in 1961, were responsible for 6% of cases of gonorrhoea in men and 26% in women. Between 1957 and 1961 gonorrhoea increased among the ages 15 to 20 by 63% in men and 78% in women, and the increase could not be put down simply to an increase in population (see Table 27).

TABLE 27

Cases of Gonorrhoea among young people aged 15–24 in England and Wales, 1957 and 1961

Year	*1957*		*1961*		*% increase*	
	M	*F*	*M*	*F*	*M*	*F*
Population aged 15–24 (thousands)	2,757	2,775	3,072	3,015	11·4	8·6
Cases of gonnorrhoea (treated at clinics)	4,999	2,316	8,142	4,123	62·9	78·0
Rates per 10,000 population	17·5	8·3	26·0	13·7	48·6	65·1

From *W.H.O. Chronicle*, April 1965.

In the U.S.A. cases of syphilis among 19-year olds and below rose from 1,179 in 1956, to 3,852 in 1962. Cases of gonorrhoea increased from 47,911 to 55,560 (see Table 28). Similar increases in V.D. among young people have been reported from many other countries.

These facts are no doubt partly due to the earlier physical maturity of young people today which can conveniently be measured by the age of menarche. Fig. 47 shows that menarche in Europe has been getting earlier during the last century by between three and four months each decade; in 1840 the average age of menstruation was 17, while in 1960 it was about 13½. The trend in boys is similar but it must be remembered these are *average* figures and there is great variation in them. Has the trend to early maturity stopped? The graphs could be carried forward for 20 years or so, then perhaps the lower limit of menarche set by a genetic mechanism might be reached.

What are the causes of this trend to earlier maturity? Better nutrition and health are probably the main causes, and height and weight are

TABLE 28

Infectious Venereal Disease in the U.S.A., by Diagnosis and Age, 1956–62

	0–9 years		10–14 years		15–19 years		20–24 years		0–19 years		0–24 years	
Calendar years	Syphilis cases*	Gonorrhoea cases	Syphilis cases*	Gonorrhoea cases	Syphilis cases*	Gonorrhoea cases	Syphilis cases*	Gonorrhoea cases	Syphilis cases*	Gonorrhoea cases	Syphilis cases*	Gonorrhoea cases
1956	11	1,222	75	2,425	1,093	44,264	1,778	74,755	1,179	47,911	2,957	122,666
1957	33	1,628	80	2,363	1,192	43,705	1,857	70,777	1,305	47,696	3,162	118,473
1958	23	1,164	90	2,706	1,228	48,723	2,005	76,964	1,341	52,593	3,346	129,557
1959	39	1,328	109	2,600	1,668	50,324	2,779	81,076	1,816	54,252	4,595	135,328
1960	20	1,619	139	3,261	2,577	53,649	4,692	87,823	2,736	58,529	7,428	146,352
1961	35	1,615	210	2,567	3,215	52,131	5,575	90,686	3,460	56,343	9,035	146,999
1962	40	1,459	225	2,422	3,587	51,679	6,063	91,588	3,852	55,560	9,915	147,148

* Including both primary and secondary cases. From *W.H.O. Chronicle*, April 1965.

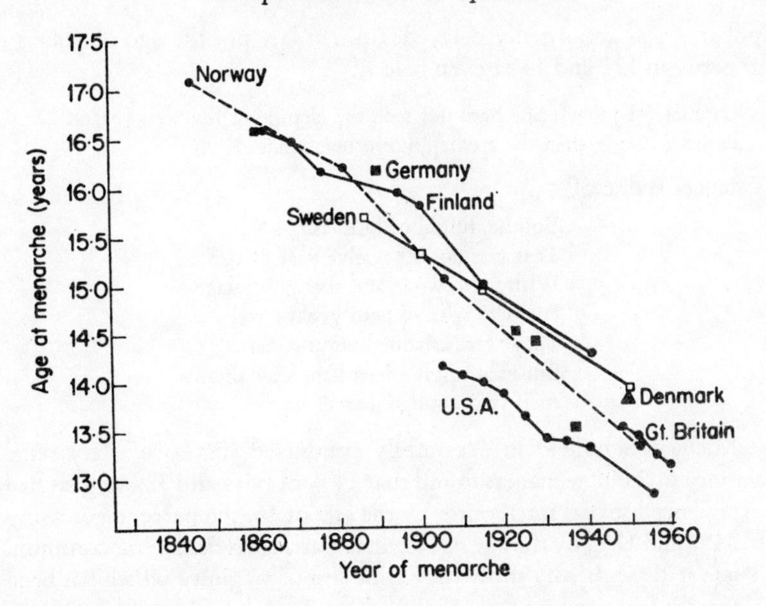

FIG. 47 A selection of the best available data showing the trend towards earlier maturing during the last century.

After Tanner, J. M., 'The Trend Towards Earlier Physical Maturation,' *Biological Aspects of Social Problems*, Ed. Meade & Parkes (Oliver & Boyd, 1965).

persistently associated with it. For example children of 12, thirty or forty years ago had, on average, the same weight as the 11-year-olds of today, and the timing mechanisms controlling menarche are set just that much further forward. Better health and feeding might advance maturity, but under-exercise and stifling clothes (the stays of Victorians) may have been important in delaying it. Besides these physiological explanations there is also a social one which affects behaviour—the emphasis on sexual behaviour in popular reading, plays and films, on television and in advertising. Hand in hand with earlier maturity the average age of marriage has fallen during the present century.

The increase in precocious sexual behaviour among teenagers needs to be looked at against three pieces of evidence. First, that early maturity may be only a return to normal after a period in the late eighteenth and early nineteenth centuries when for some reason puberty was abnormally delayed. Roman law put the age of puberty at between 12½ and 14 or even below:

> Capulet (of Juliet): She hath not seen the change of fourteen years. . . .
> Paris: Younger than she are happy mothers made.*

Chaucer spoke of

> Youthe, fulfild of lustinesse
> That was not yit twelve year of age
> With herte wilde and thought volage . . .
> For who spak of hem yvel or wel
> They were ashamed never-a-del
> But men mighte seen hem kisse there
> As if two younge douves were . . .†

Michael Schofield in a carefully conducted survey of the sexual activity of 2,000 teenagers found that 14% of boys and 5% of girls had experienced sexual intercourse by the age of 16, the percentages rising to 34% and 17% by the age of 18 when marriage is becoming common. These statistics hardly show the extinction of virginity which has been gloomily forecast by some. Eighty-four per cent of boys and 82% of girls, according to Schofield had some knowledge of birth control but 'the majority neither took precautions themselves nor insisted on their partners using any'.

Table 29 shows that an important inhibiting influence for not having sexual intercourse was fear of pregnancy but the most usual restraint for girls and an important restraint for boys was moral reasons. This is not surprising; for generations, as Dr Alex Comfort writes, conscience, the unwanted baby, fear of God and the spirochaete have been firm allies in saving us from the flesh.

Knowledge of V.D. among teenagers was found by Schofield to be sparse and many fears and old wives' tales still persisted among them. About half would not be able to recognize symptoms if they became infected.

Third, that despite the impression given by the newspapers the overall increase of illegitimate births and of pre-maritally conceived children has remained stable since the 1930's but with earlier maturity the distribution has shifted to the younger age groups. As Alex Comfort has written 'the magnitude of the shift is such that the interests and problems of the sixth former today are roughly those of the undergraduate yesterday'.

* Romeo and Juliet, I, ii.

† The Romaunt of the Rose.

TABLE 29

Reasons for not having sexual intercourse classified into eight categories. From Michael Schofield: The Sexual Behaviour of Young People, *Longmans, 1965.*

Reason	*Boys*	*Girls*
	%	%
Fear pregnancy	24	17
Moral reasons	19	40
Religious reasons	3	5
Girl's reputation	11	9
Keep virginity	1	4
Fear V.D.	1	2
No reasons	14	2
Don't know	27	21
Total	100	100
Number in Sample	596	769

VENEREAL INFECTIONS

The British Medical Association in 1964 said about V.D. in young people that:

> In spite of new drugs and improved methods of treatment an increase in sexually transmitted disease has taken place, and this most marked among young people. . . . The actual increase may well be 15% greater than the official records show. This increase argues that there has also been an increase in promiscuous sexual behaviour. . . .
>
> It is obvious that the problem has medical aspects, but we believe it to be primarily a social and moral problem. . . . There are signs that a 'fashion' (of promiscuity) may well be becoming current, but it would be both hazardous and unjust to ascribe this to one group of people in particular. . . . The increase in promiscuity among young people is a symptom of change in society as a whole for which all sections of the community are responsible.

A report published by the World Health Organization summarizes a number of reasons for the increase in precocious sexual behaviour among young people. Besides the trend to early maturity and the emphasis on sex in advertising, films and popular reading, other reasons given are:

Ignorance of the nature and meaning of sex and of the dangers of abuse of the sexual function;

Decline in religious faith;

Lack of discipline in home life and of parental supervision, which is one of the symptoms of the disruption of family life.

The disappearance of the chaperone. Lively, unsupervised youngsters who have a desire to be 'with it' are more likely to sample sex, and as the report states, freeing young women from submission to the supervision of chaperones has led to a measure of licence. Young men and women quite often meet in places where they can get a drink and sometimes too much drink is taken and self-restraint relaxed. In one survey about 77% of men and 25% of women had taken alcohol before the intercourse that led to their becoming infected with V.D.

Whether or not the actual population incidence of syphilis has increased in comparison with what is known of V.D. in cities in the eighteenth century is a matter for debate. Its methods of transmission have certainly changed since then when it was spread by organized prostitution. In the twentieth century it has been propagated by war—conquering armies, defeated populations, populations undergoing the strain of war. Now its spread is by a changed society which is prosperous but aimless. As far as young people are concerned Michael Schofield writes:

the problem of venereal disease among teenagers is not so great as the problem of illegitimacy. Nevertheless there has been an increase in the number of infections in the last few years, although this increase has not been so great in this age group as the rise among other sections of the community. The main increases in gonorrhoea have been among immigrants, homosexuals and young adults aged twenty to twenty-five. It is this age group that has given rise to misunderstandings because in the past, annual figures for infections have been grouped into age ranges, starting with fifteen to twenty-four.

Once more as with the alleged increase in neurosis it seems as if the morals of teenagers now are no 'worse' than they were in the past. 'In fact, so far as there is a "problem" of teenage morals it is a reflection of the earlier physical maturity of young people—their confrontation with moral choices at an earlier age.'

27: Old Age

I, a stranger and afraid
In a world I never made.

A. E. Houseman.

VIOLENCE and sex are two manifestations of teenage behaviour that hit the headlines in our society. Besides these, rates of suicide among adolescents, drink offences and road accidents have increased, and this trend is not restricted to Britain but applies to the western world generally. The other end of life is perhaps more peaceful but this too can be spoilt by ill-health, for old age brings with it inevitable enfeeblement, degeneration and death, and more cruel than these, it can bring isolation and loneliness which are not inevitable.

It was pointed out in Chapter 19 that the commonest age of dying may soon shift from 75 to 85 if there is a real break-through in discovering the causes of aging. Now, because of healthier living and advances in medicine, one person in every nine in Britain is 65 or over. At the turn of the century the figure was one in every 21 and by the late 1970's it is likely to be one in seven—about eight million people compared with about six million now. Similar trends are found in all rich countries but Britain shows two special characteristics: the rate of increase in the number of old people is faster than in many other countries and there are more elderly women than men.

As people become older they become more and more dependent upon others to care for them, and while most over-65's are in regular contact with close relatives, about a million live alone and are isolated. A study in Bethnal Green in East London showed that 70% of old people had regular help from relatives, 20% had some help, and 10% had none and were more or less isolated socially. Many of the lonely are single women or childless widows or widowers. Many of them cannot get about—they have difficulty in walking, getting up and down stairs, washing and dressing—or are bedridden. Studies of the health of men over 70 in Birmingham showed that there was not always a consistent increase in illness with old age. This is because those who are ill go to hospital or die and the studies therefore were related to the fittest who survived. A high rate of bronchitis, high blood pressure, coronary artery disease, arthritis, hernia and peptic ulcers was recorded, but none of these increased regularly with age except arthritis, which does not kill. After 75 most elderly men were not able to walk far and after 80 the majority were house-bound. This is due to failures in heart, lungs and brain, a deterioration in muscle power and joint movement and bad feet. Many of these old people cannot get out because of severe difficulty in seeing and hearing and a fear of negotiating traffic.

Professor Peter Townsend has calculated that about one million people over 65 need expert help in looking after their feet, about 750,000 are hard of hearing, and about 500,000 have severe difficulties in seeing, while about 700,000 suffer from severe incapacity for movement. Such problems of old age grow each decade—for example there are 40% more 80 year olds now than there were 10 years ago.

SUICIDE

It is not surprising that the suicide rate among the elderly rises; many of them have their world cut suddenly away by retirement and they become socially isolated. Social isolation is the most common factor among the causes of suicide. As Professor Erwin Stengel* writes: 'in Western society the large family group with three or four generations no longer exists. Today more people live alone than ever before. When illness or other misfortunes hit them suicide becomes a real danger.'

* *Weekend Telegraph*, Oct. 14, 1966.

This fact is borne out by a study of the suicide rates over a number of years in two industrial cities in the North of England. Usually the incidence of suicide is higher the denser the urban population. However the opposite was recorded in city A whose suicide rate was greatly in excess of that of city B, although A's population was only one-fifth of B's. Moreover their populations were similar in class structure; important because in western societies the suicide rate is higher among the professional, managerial, and business people than among working classes. But the *age* composition of the two populations differed greatly as Table 30 shows.

TABLE 30

Suicide rate

	A	*B*
1952	19·1	8·0
1953	27·6	9·3
1954	30·1	5·6
1955	26·5	9·0
1956	42·5	10·0

Percentage of people aged 60 and over

	Male	*Female*
A	21·6	20·0
B	13·4	16·7

Annual death rates per million

	Male	*Female*
A	16,405	14,087
B	13,312	10,670

From Facts, 'Figures and Suicide', by E. Stengel. *Discovery*, 1964

Although there might have been other factors at work the larger number of old people in A compared with B made the suicide rate of the former excessive. Professor Stengel who studied this problem stated that there was evidence that the population of A had become 'over-aged' because the young people had gone elsewhere as local industry declined. The high suicide rate in West Berlin (37 per 100,000 population as against 19 in London) is probably due to the large number of old people there, the young having continually emigrated. We must expect therefore higher suicide rates in aging populations and in local populations such as towns and cities that have lost their trade and in which old age groups persist.

The nutrition of the elderly naturally depends on income and help;

Townsend calculated that about 400,000 were not getting properly fed because they were having difficulty with cooking and feeling the need for having meals brought to their homes. Nevertheless, as a *single group*, old age pensioners compared with other sections of the population were not under-nourished on the evidence of the results of National Food Surveys from 1950 to 1960. This result is curious for it is at variance with other evidence on malnutrition among the elderly. For example, in 1955, 20% of old people examined in Sheffield were found to be in a state of poor nutrition.

Most sociologists believe that old people should not be isolated as a special group, but should be integrated with the rest of the community they know, preferably in their own homes rather than in institutions. It is recognized that they need help (and this some are getting through Home Helps and the meals service) so that they can maintain their independence for as long as possible as useful members of the community.

So often people make the elderly *feel* old. A century ago a woman of 40 put on her mob cap and settled down to old age. Even 30 years ago a woman of 40 who wore bright clothes drew the remark from many that 'mutton was dressed up as lamb' and great grandma put on jet and button boots and black clothes at 50. Nowadays the climate has changed and it is laughable to think of these attitudes. As Comfort writes, 'we make people socially old by retiring them; we may even, by the same token, make them *physically* old for mind, body and society interact in a degree that can still amaze us'. Ecologists have realized that to divorce the individual organism from its background is unreal, the organism and its habitat are often necessary to each other. So it is with men. Keeping active with hobbies and work, having the respect of our fellow species, makes us live longer and loss of these intangibles makes us die young. Primitive societies may be able to teach us something here. The relatively few old people in them are regarded as 'repositories of knowledge, importers of valuable information and mediators between their fellows and the fearful supernatural powers'. By doing this primitive societies succeed in providing the social conditions to use the abilities of old people.

As to the physical side of aging, physical performances of aging men show that it is not necessarily muscular weakness that makes work too 'heavy' for them (although old muscle is apparently less efficient at getting energy out of sugar) but having to hurry. 'Lighter work' means work that can be carried out at a slower pace. Mental performance, that is speed of thought, learning power and originality, deteriorates with age in all but a few. To make up for this there is an increase in the care taken over work and in experience.

All the evidence points to the need for helping old people to maintain their dignity and status in the community by a special effort to provide suitable work for them. Our own society supports elder statesmen, judges, clergy, gardeners, and farmers in prolonged and useful activity and by the same token it might extend this social support and respect to many more people past 60.

Nevertheless it is obvious that with the increasing life-span of man a price has to be paid. Newton completed his greatest work by the time he was 25. Most great poems are written at about 30, most great novels at around 45. The vitality of the Winston Churchills and Bertrand Russells (and the remarkable Georgian centenarians—see Chapter 19) is no doubt due in part to their genetic endowment which helps to determine the life-span. It is also due to culture and education, and the support of their peers. People who use their minds a great deal seem to retain their intellectual capacities to a much greater extent than those in jobs which make few intellectual demands. But most of us degenerate more quickly than the two giants named and medical advance props us up. In fact it keeps people alive who a century ago might have died at 35 of pneumonia but who now live on to get arthritis or stroke in old age. Most people prefer this and the community benefits.

Some biologists have asked whether it would be possible to pick out, in babyhood, people who genetically are likely to die of stroke or chronic bronchitis in their 60's or 70's. At present, however, it is impossible to say which organs of a child will break down in old age, and even if this could be done it is doubtful whether life could be patched up, for other organs by that time might be on the verge of a breakdown. And, for the population as a whole, early detection in a baby of organs likely to cause trouble 60 or 70 years hence would not be useful since the baby by that time would have survived to reproduce and perpetuate the trait.

A number of people feel strongly that anyone who is of sound mind should, before he becomes chronically ill and feeble-minded, make a declaration to be lodged with the family doctor stating that apart from pain-killing drugs no other action to prolong life should be taken. This philosophy echoes the words of Clough:

> Thou shalt not kill, but need'st not strive
> Officiously to keep alive.

Indeed, although man dreams of living forever, it is not possible (except in the faces and actions of his children) nor is it desired by most people. Death is a check to over-population and a defence against the intolerable prospect of immortality. Gulliver when speaking of the immortal Struldbruggs in the kingdom of Luggwagg gave us a clear and candid description of old age which is enough to extinguish any ambition (for most of us) of reaching the century.

At Ninety they lose their Teeth and Hair; they have . . . no Distinction of Taste, but eat and drink . . . without Relish or Appetite . . . In talking they forget the common Appellation of Things, and . . . for the same reason they never can amuse themselves with reading, because their Memory will not serve to carry them from the Beginning of a Sentence to the End. . . . and by this Defect they are deprived of the only Entertainment whereof they might otherwise be capable.

They were the most mortifying Sight I ever beheld; and the Women were more horrible than the Men. Besides the usual Deformities in extreme old age they acquired an additional Ghastliness in Proportion to their Number of Years, which is not to be described.

And Gulliver concludes, 'the reader will easily believe that, from what I had heard and seen, my keen Appetite for Perpetuity of Life was much abated'.

Part 6: Brain and Behaviour

Knowledge of the activity of the 1,300 g. of brain packed into the skull is far less than that of outer space, yet man's unique nature is based on his brain. His ability to enjoy a fine sunset, or appreciate a decent meal or recognize a fine painting, his doubts or hopes about himself or his family or the world situation and his ability to put these manifestations of consciousness into words and pass them down the generations by speech or books is characteristically human and is based on the activity of about ten thousand million nerve cells in the cerebral cortex of the brain. There are as yet very few established facts about how this 'enchanted loom' works. We know that its function can be imitated by machines, that its working can be compared with physico-chemical processes, but that it remains unique and intractable. We know that it consumes sugar and oxygen avidly, that it is quickly choked into inactivity and death by carbon dioxide which has to be shifted away quickly, that it works on the quantity of electricity necessary to keep a rather dim light bulb going, and that the study of brain function has tended to emphasize the importance of heredity. These rather gross facts about brain function however do not help to explain the subtle changes of mood that flesh is heir to. Drugs, as we know, affect mood and behaviour, but there is only the flimsiest notion of why they should do so. Watching animals behave does not give us much of a clue about human behaviour. The human machine is much too complicated to bear comparison with most animals. It is rather like trying to compare a jet aeroplane with a rubber-driven toy. And man can lie about his behaviour while animals cannot. What can help, as we shall see, is using the objective methods and language of ethologists to study man and using such studies to suggest hypotheses about human behaviour. But if we are to get very far in knowing more about ourselves we shall have to swallow our pride and allow ourselves to be subjected to scientific study.

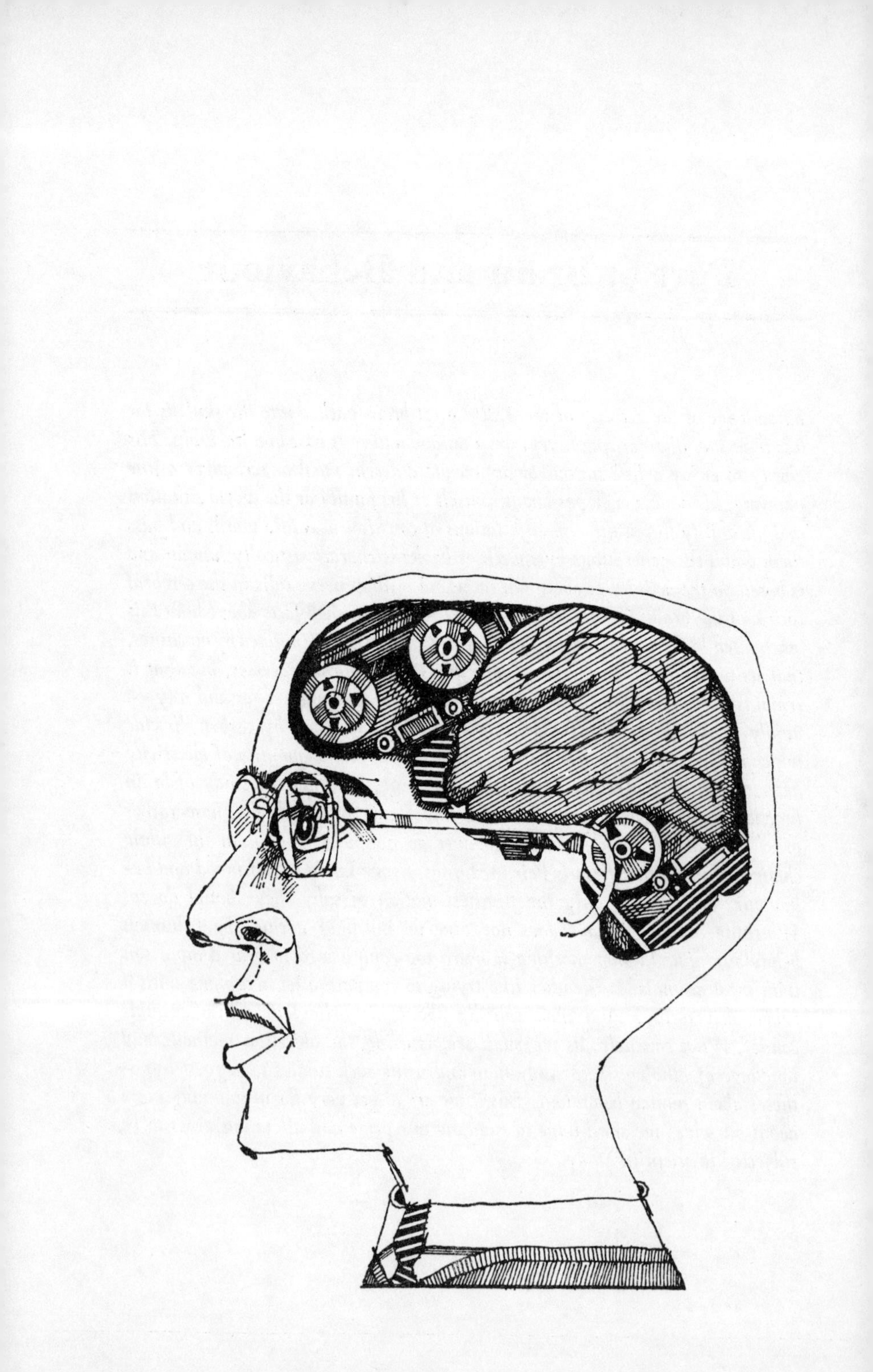

28: The Physics and Chemistry of the Brain

MAN, (so we believe) is set apart from other animals by such characteristics as speech, his strong sense of religion, his moral sense, his 'bloody-mindedness', his imagination and his appreciation of beauty. These qualities which lift him above the beasts manifest themselves in his behaviour which is a function of the brain. We all know that we learn about the world through our ears, eyes and sense of touch, and we know, for example, that when light falls on the retina, messages of an electrical nature are sent along nerves to the parts of the brain that are known to deal with sight. An accurate copy of the distribution of light on the retina is not passed along nerves but rather a simplified code which somehow the brain turns into a picture. As Russell Brain wrote so concisely 'the only necessary condition of the observer's seeing colours, hearing sounds, and experiencing his own body is that the appropriate physiological events shall occur in the appropriate areas of his brain'. This is all rather unsatisfactory and vague but we are as yet in a very primitive stage in our understanding of how the brain works or how it affects behaviour.

This is not only for the obvious reasons that its machinery is extremely complicated and delicate and hidden, but also less obviously because a language for describing how the brain works and 'what it is for' has not been easy to find.

BRAINS AND COMPUTERS

Fortunately with the development of computer engineering a language has been found and analogies have been made which have enabled brain function especially to be studied fruitfully. Electronic calculating machines can now be built which can be used for storing, transmitting and manipulating information. Some can even be made which behave in ways unpredicted by their designers; an element of random choice can be built into these machines so that their final behaviour cannot be forecast with certainty—indeed, they appear to have a will of their own. The most useful of these machines, from the point of view of brain analogies, are those whose rules of behaviour are adjusted according to the degree of success they achieve by trial and error and they do this by adjusting the relative probabilities of different patterns of activity. The brain likewise somehow stores information in code and is able to calculate the probable outcome of an action by comparing new information against encoded past experience. If the comparison is valid an accurate forecast results. This ability of man to learn so that his patterns of behaviour will deal more effectively with the future environment is a characteristic feature which cuts him off from the crude machine analogies of the past.

One obvious difference between computers and brains is that of sheer size. For efficient storage of information in brains a large number of short nerve cells seems to be required—the grey matter of the human brain has 15,000 million of them, packed into a sheet of about 2,000 sq. cm in area and about 3 mm in thickness. Fifteen years ago when J. Z. Young gave the Reith lectures, mechanical calculators had twenty or thirty thousand valves and one with as many valves as there are cells in the cortex would need, as Professor Young said, a huge building to house it, while all the waters of Niagara would not be enough to work and cool it. Transistors have now replaced valves but mechanical imitations of the brain are still a long way off. For one thing it is easy to bark up the wrong tree by making more and more machines which behave superficially like the brain. The most useful is that which is made to test some hypothesis about brain function. Its working is compared with that of the living brain, so that the differences between it and the machine will point the way by fruitful hypotheses to new and more accurate machines to be tested once again against the brain. By these methods machines and the brain will become more alike and if eventually a machine can be made which could, for instance, read human handwriting, then an important step will have been taken in understanding how the brain works. Models of the brain need to be made not only to function like a 'normal' brain but to imitate brain disease too.

BRAIN CHEMISTRY

Comparison of brains with computers suggests that the brain might be very much like a complicated electronic machine, but really it is much more of a *chemical* instrument than a physical one and its functioning can be affected by other chemical substances—drugs and other chemicals like alcohol. If alcohol was poured over a computer it would not affect its rate of operation but its effect on a chemical system such as the brain is profound as we know to our cost.

A good deal of information has been built up over the years about the chemical processes that govern the simple supply of power to the brain for its use in more complex ways. The raw materials for the power or energy are oxygen and glucose and the brain's need for these is continuous and avid. Glucose, at the rate of one teaspoonful each hour, is the basic energy food of the brain and oxygen is needed to make the sugar yield up its energy. Oxygen is consumed by the whole brain at about 46 ml per minute which represents about one quarter of the total oxygen used by the resting body, but individual parts of the brain, for example, the visual centres when working, take in oxygen at a greater rate than other parts. The breakdown of glucose by oxygen yields about 20 watts—the electricity that is needed to feed a very dim light bulb—yet this little power supply enables a genius to create or a maniac to kill besides allowing the basic patterns of sleeping, talking or walking. How different from the thousands of watts needed to keep a computer going to perform its relatively restricted operations!

To cut down the power supply to the brain by reducing the supply of oxygen or sugar can affect the mental state. The interference of the power supply to the brain may be brought about by drugs or chemicals or perhaps by old age. In an alcoholic coma the power supply to the brain is cut in half. But the odd and interesting fact is that in severe mental illnesses such as schizophrenia, the power supply to the brain is normal taking normal as 100%.

TABLE 31

*Oxygen consumption of the brain and mental state**

Condition	*Percentage of Normal*
Normal sleep	97
Schizophrenia	100
Mental arithmetic	102
Anxiety	118
Alcohol coma	49
Anaesthetic	64

* From Kety, *Control of the Mind*, ed. Farber, McGraw Hill, 1961.

We can conclude that it takes just as much oxygen to think a queer thought as a normal one, and even when the brain is using more oxygen than normal as in an anxiety state, the quality of thought is not improved. Eating more sugar or forcibly increasing the oxygen supply to the brain will *not* better the quality of the thinking any more than putting more current through a T.V. set will improve its performance. Sleep uses the same amount of power as being awake. There seems merely to be a redistribution of energy, utilizing different circuits in the brain—and its sleep differs from the deepest sleep brought about by an anaesthetic.

The crude power supply to man's most unique organ certainly acts on the brain as a limiting factor to general activity, but it is not a *determining* factor in behaviour. Something more subtle is necessary to channel the crude power into finer shades of mood. This is discussed in Chapter 29.

It is essential in seeking to understand the brain that collaboration should exist between computer engineers, physicists and biochemists on the one hand and psychiatrists and brain surgeons who deal with the living brain, on the other. In between lie the essential links, the experimental biologists, who are learning much of fundamental importance about the brain through the study of behaviour of the octopus and cats. Collaboration between such teams has given us some rather slender information as to how the brain might work.

A MODEL OF THE BRAIN

One important hypothesis suggests that the brain contains a model of the outside world. We are so familiar with this model that we think it *is* the outside world, but what we are really aware of is an imitation world, a tool which we can manipulate in the way that will suit us best and so find out how to manipulate the real world which it is supposed to represent. The model or picture, call it what you will, is a highly personal one, because it is made up of data which are *selected*. That is, in our daily lives, indeed from childhood, we all tend to ignore some things and emphasize others. If the model is constructed from data supplied mainly through the eyes, ears and sense of touch (and it must be) then it is clear that faulty sense organs will produce a model that inadequately represents the environment. Perhaps some of our great artists may have had eye defects; Goya's elongated figures could be accounted for by such a defect. When people have been blind from birth and their blindness is cured they see nothing but confused images which they are not able to identify; these people very gradually learn by trial and error to make a practical model. The practical usefulness of the model is that it agrees very closely with the outside world. When we cross the road and avoid traffic we are really dodging the moving buses and cars in the mind. If the model is inadequate or is used for wishful thinking we should soon be in trouble.

THE MEMORY ALPHABET

What is the physical basis of this model? Is it made out of anything in the head? It is possible that it is assembled out of units which are inherited and which must be in the brain. These basic units, which may be compared with a ready-made alphabet, may be in real terms the cells of the brain, which are capable of keeping account of events and remembering things in almost unlimited number. These cells are not all the same and their particular character is given to them by the receiving gear in each cell. This is formed by thin branching tendrils, the dendrites; some cells have few unbranched dendrites and others have many, and thus it is possible that the differing shapes of the dendrites form the 'memory alphabet'. In animals such as the guinea pig and the mouse the dendritic patterns are fixed very early, sometimes at birth. In man they may be fixed but it is more likely perhaps that the dendrites can to some extent respond—that is alter in shape—to experiences during life-time. Sir John Eccles believes that the richness of human performance stems from the potentiality of the human cerebral cortex for developing 'subtle and complex neuronal patterns of the utmost variety', a potentiality perhaps not shared by even the most intelligent animal.

The basic requirement of a dendrite-alphabet is that it is a suitable one to store representations of all the situations an animal is likely to meet in a lifetime. Evidence from experiments on the frog's eye shows that its 'visual alphabet' is simple and allows storage of information that is likely to be of interest to the frog. Its visual alphabet might consist of only six 'letters' which can code experiences such as tracking moving insects or jumping towards blue rather than other colours—that is, they jump towards open water rather than towards vegetation. Our own visual alphabet must be a good deal more complicated than this.

More light is being shed on this 'pre-set' visual alphabet in humans by the work of an American infant teacher in California and two German scientists. The teacher has analysed 300,000 drawings and paintings from two-year-old American, Chinese, French, English and Negro children and has concluded that they are built up from 20 basic scribbles and six typical diagrams (see Fig. 48). The child later combines them into more complex forms built up from two or more diagrams and this enables him at about the age of four to draw such things as the sun, people, houses, animals and flowers. The two scientists have been investigating luminous patterns seen when the eyes are closed and when the brain is stimulated electrically. Copies of 520 of these 'phosphenes' observed by 313 people have been collected and analysed into 15 groups. Ninety per cent of these phosphenes can be found among the 20 basic scribbles of children although they were collected independently. Both have in common arcs, crosses, waves, lines, combined patterns, circles, dots, odd figures, quadrangles, spirals, poles, triangles, and 'cherries'. Lattices and fingers do not appear very often in the scribbles. It looks therefore as if some kind of visual alphabet exists in the nerve cells of the brain which is laid down early and is common to all men. These nerve networks need to be activated by scribbling in children.

The memory alphabet might be used as a model against which new information, fed in through the senses, can be matched. Before a decision is reached and action taken small scale experiments within the bounds of the model go on in the brain. When a forecast has been reached which does not conflict with any rules of conduct the animal (or ourselves) has learned through experience, then action or decision follows. This is what many computers can do as previously described. An animal might alter its behaviour in accordance with changing

INFANT SCRIBBLES AND PATTERNS IN THE BRAIN

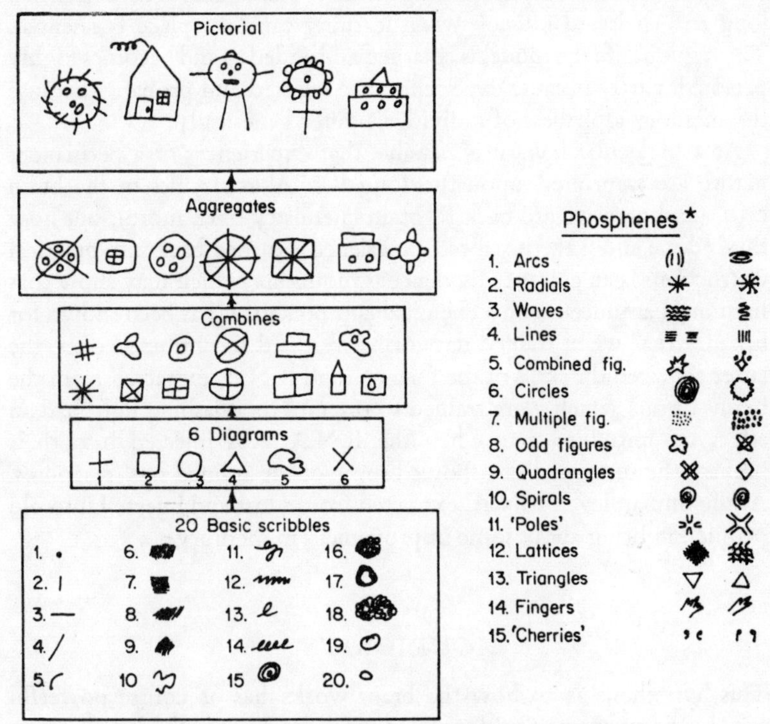

FIG. 48 Development of outlinings of 3-6-year-old children.
After Kellog, Knoll and Kugler, *Nature*, 11 December 1965.
* See text.

external conditions so that it keeps itself at the right temperature (in a 'steady state'). A man may do the same, for example when he is driving a car. His brain forecasts his position in relation to other moving traffic in his mental picture and his life depends on the accuracy of his forecasts.

The picture or model of the world is not a static one. A baby has no model—only its inherited brain-alphabet from which it will 'construct' its own personal model of the world. By staring, feeling, tasting, it will slowly make a 'construct' of its world and learn how to use its muscles within the framework of its picture of the world. Learning at this stage is slow and inefficient, but simultaneously with the building of its mental picture the baby is somehow organizing the huge mass of connecting nerve cells in the forebrain so that later on rapid responses and rapid learning can take place. This is why in man a long and sheltered infancy when learning can take place is essential. Throughout life the model is enlarged and added to and becomes highly personal, partly because experiences are selected and probably because the memory alphabets of individuals differ genetically.

At a molecular level it is possible that experiences of a permanent nature are imprinted upon the long R.N.A. molecules in the brain cells, (and here we are back to brain chemistry once more), but how this is done and how the stored experience is turned back into practical instructions is an enigma. Evidence is mounting which may show that memory has indeed such a chemical component. It has been shown for instance that when trained flatworms were fed to untrained ones, the latter acquired the skills of the former. And R.N.A. extracted from the brains of rats which were trained to respond to a flashing light had an effect on untrained rats when the R.N.A. was injected into their brains—the untrained rats showed some of the trained rats' responses. At the human level, R.N.A. extracted from yeast and injected into old people can bring about some improvement in memory.

CURIOSITY

This hypothesis as to how the brain works has of course powerful social consequences. The dogma or propaganda or orthodox thinking which we may be exposed to day after day in our everyday life (especially our early life) can cripple minds with imposed attitudes and 'establishment' thinking, so that the picture of the world is warped, stunted and inadequate. To gain a useful model we must always try to understand and make sense of the impressions we get from the senses. We need to notice and try to explain why some bits of experience do not fit in with our model. A consequence of this is curiosity. The human brain works best in an environment which provides the stimulus of meeting new situations and surmounting difficulties. Studies of the behaviour of men in isolation have shown the need for constant novelty if normal behaviour is to be preserved and this is well known in animal studies. Rats sometimes appear to behave as if they were easily bored and are driven to seek new experiences. Intelligent children too, as Bertrand Russell once remarked, spend much of their time in a state of boredom because their environment is unchallenging. And the importance of novelty of experience in the development of curiosity and intelligence must make us pity the baby

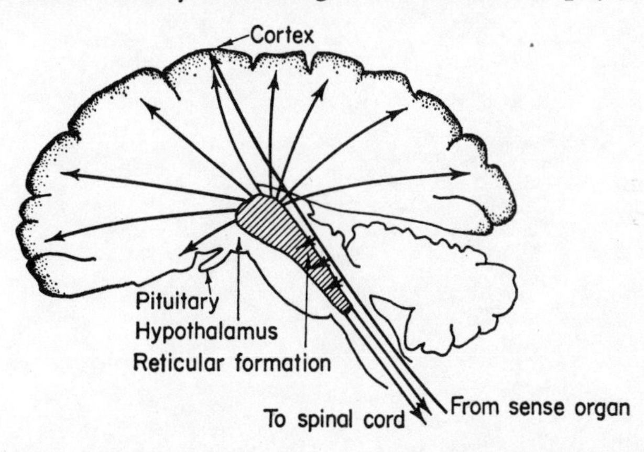

FIG. 49 A diagram of the brain. At the top is a cerebral hemisphere with its outer lining or cortex of grey matter, beneath are the mid and lower brain leading down to the spinal cord. The cortex receives messages by pathways which relay impulses direct from sense organs (the heavy arrow). The messages also branch off into the reticular formation and help to keep it excited. The reticular formation itself gives off streams of nerve impulses—some pass up to the cortex, others forward to the hypothalamus, while others flow down the spinal cord to nerve cells which control muscles and other bodily organs.

After Oswald, I., 'Sleeping and Dreaming' in *Science Survey B* (Penguin Books), 1965.

who is made to sit in his pram for long periods! Why do we seem to become less curious as we grow up? The brain machinery might still be as good but probably the urge to explore fades.

The part of the brain which is the pacemaker to the rest and keeps the grey matter (cortex) pepped up to cope with new situations is called the reticular formation (see Fig. 49). During wakefulness the reticular formation, which is deep within the brain but is connected by a profuse network of nerves to other parts of the brain and spinal cord, sends out messages to the cortex and keeps it wide awake and curious. It also keeps muscles toned up and the breathing stimulated. And the more stimuli that are arriving from the outside world and from the brain itself the livelier the action of the brain pacemaker and the more receptive is the cortex. The accuracy of our model of the world depends to a large extent on our curiosity and this perhaps on the state of the reticular formation, and this on the crude supply of power to the brain.

29: Doors in the Wall

ALDOUS HUXLEY wrote in his book, *The Doors of Perception*, that he doubted whether men and women would ever be able to give up 'Artificial Paradises' created for them by alcohol, drugs and smoking; 'life is too grim and uncompromising', and escape from it, if only for a few moments, has always been one of man's principal cravings. 'Art and religion, carnivals and Saturnalia, dancing and listening to oratory—all these have served in H. G. Wells' phrase, as "Doors in the Wall".' And no civilization has failed to discover alcohol as a mind-easer, while tobacco has brought relaxation to Western man for four centuries. As well as the 'legal' drugs there are countless others, some the products of berries, roots, cacti, leaves and bark which have been used from time immemorial to ease man's primitive fears or to produce mystical states during religious rites and others that are man-made. These have poured in abundant variety from the drug industry in the last 15 years; stimulant drugs to increase wakefulness and reduce fatigue; anti-depression drugs which not only help to fight off the black depression of the mentally ill but speed reaction time and make words come easier to the lips; major and minor tranquilizers that calm the mind without confusing it or sending it to sleep. Apart from these synthetics there has been a rediscovery of the hallucinogens (e.g. mescaline) which can mimic in normal healthy people the picture of schizophrenia, a major mental illness which may have produced the visions of Blake (as a boy he described a tree which he saw filled with angels), Van Gogh's glowing cornfields, Bosch's minutely-detailed monsters, and perhaps the brilliant, exploding colours seen by other painters and mystics not only with the inward eye but apparently in the solid world around them. Aldous Huxley, after taking mescaline, reported that he saw flowers and even the folds in clothes with enhanced beauty and richness. Perhaps there is something in the view of certain psychiatrists that schizophrenics in other civilizations than our own would be treated as individuals who had achieved a higher state of consciousness.

There may indeed be a 'need', as Huxley wrote, 'for frequent chemical vacations from intolerable selfhood and repulsive surroundings'. To achieve this Huxley would like to see the production of a synthetic non-toxic 'escape' drug, potent in minute quantities and not having the undesirable social consequences of alcohol nor the dangers to heart and lungs of nicotine. But most drugs have been used, in fact, to control abnormal moods in mentally ill people and the anti-depressants and the major tranquilizers have been largely responsible, as we have seen, for the big increase in discharge rates of mental patients from hospitals in recent years.

Some of these drugs are being peddled in thousands to teenagers for kicks and excitement. Perhaps 50,000 teenagers have taken amphetamine-barbiturate mixtures (purple hearts) in the London connurbation alone according to a recent report in *The Times*. Most of the drugs are stolen from chemists' shops and sold in clubs and cafes. A purple heart tablet may now cost as much as one and sixpence. Before the unauthorized possession of amphetamines was made an offence they could be had for sixpence each. Kicks and excitement are the 'Doors in the Wall' for these jaded adolescents, but eventually the pills reduce initiative, vigour and drive as well as having harmful side effects.

BRAIN CHEMISTRY

It would be of interest to know how these drugs work to control mood or produce visions. Unfortunately it is not known how they affect the brain/mind/behaviour complex—perhaps one result of the failure in collaboration between biochemists and psychiatrists in the past. Recently, however, a number of ideas have been put forward. An obvious one is that the drug itself is supplying some chemical which the brain, through faulty chemistry, cannot make, thus causing abnormal mood and behaviour. A case in point is that of mescaline. By a stroke of genius a young Canadian psychiatrist noticed a similarity between the chemical structure of mescaline and adrenaline, a hormone which is released from a gland over the kidney when the body is alerted for 'fight or flight'. A possible cause of schizophrenia, therefore, may be a faulty metabolism of adrenaline by the body so that a mescaline-like substance is produced which causes symptoms of schizophrenia. The search for such a chemical substance or substances has not as yet been altogether successful, although biochemistry is likely to lead to the right answer. It has in fact yielded a valuable clue. A mescaline-like substance, known as 'D.M.P.E.'* is secreted in the urine of about 60% of certain types of schizophrenics. The presence of D.M.P.E. gives a pink spot on porous paper when the urine is analysed by paper chromatography. The same test with a control group of 300 people without the disease produced only one pink spot in the whole sample. It is possible that some types of schizophrenics excrete D.M.P.E. because they simply do not have the enzymes, and therefore genes, to break it down. Bio-chemical mistakes as we have seen can lead to mental disease in phenylketonuria.

Where are these chemicals released in the brain? Research is now centred upon the hypothesis of a central 'mood regulator'—the 'limbic system'. This is a number of complex structures in the brain. The limbic system† is rich in blood vessels and is profusely cross-webbed with nerves to other parts of the brain. Most interesting, it is the site of production of chemicals which may affect mood. Certain anti-depressant drugs increase the concentration of one of these substances, serotonin, in the brain of animals, possibly by the blocking of enzyme chains, while another drug, reserpine (which can cause depression in humans) causes serotonin to diminish in concentration. Chloropromazine too, a drug which can control schizophrenia, is known to affect the lumbic system. Might this drug help to reduce a D.M.P.E.-like substance secreted in the limbic system?

The minor swings in mood that flesh is heir to in day-to-day life may well be due therefore to very slight changes in the concentration

* 3,4-dimethoxyphenylethylamine.

† An important part of the limbic system is the hypothalamus. There is some evidence that this is important 'emotional centre' in the brain.

and balance of chemicals secreted by the limbic system. This interesting notion leads to a fundamental question of great social importance. Could the behaviour of populations ever be controlled by mass use of mood regulators? There are a number of precedents for this. Alcohol was used to destroy and weaken the will of certain Indian tribes, and oriental despots have promoted the use of opium by the masses for similar purposes. But it is doubtful whether food could be salted by chemicals to soften up people and make them do as some dictator bids. In any case it can be done already to populations by voice and language and by the appeal of race-myths. We have all seen the clever-tongued huckster working spells in markets, and encouraging people to buy things they do not really want.

EDUCATION

Inevitably science will be able to control human behaviour in a gross way through advances in knowledge of brain chemistry and the effect on it of drugs. It is very doubtful however whether drugs will be able to establish *new* behaviour because this is determined largely by the information stored in the brain and no known drug can modify stored information in any fundamental way. A drug can erase certain information or suppress it, but it cannot manipulate it. And no drug can improve the mind to make it more brilliant; just as no drug can improve on the action of the normal heart or kidney. Even if a chemical could be made which could increase people's suggestibility, it is likely that the coarse-grained sense, the so-called bloody-mindedness, of the man in the street would act to check it.

No drug can remove forever critical habits of mind formed by a good education which basically forms an attitude that knows when something is proved and when it is not. And if it is not proved the mind should know how to investigate the truth of the matter by asking the right questions. Respect for evidence and a critical, questioning approach to statements made in the newspapers by politicians or anyone else are our mainstays against manipulation.

The 'drug' that can affect the mind is unquestioning obedience in thought and action (naturally necessary in some emergencies) so that evidence is no longer weighed. A too obedient child, as Brock Chisholm says, will tend to develop into a man (or woman) with lack of confidence in his own thinking and a need for direction by authority. He will be happy only when he is in a peck-order situation where he knows his place and he will tend to submit to bullying by superiors and in his turn to bullying his inferiors. His mind is easily changed from one authoritarian system to another. And in ordinary life an obedient child is 'likely to become a party hack, the obedient servant of a political machine content to leave the thinking to a "boss" and to vote or smear or riot as he is told to do'.

This kind of non-chemical thought control has been known in some degree throughout history. The torments of the Inquisition are familiar; its common feature was an alternative between violence (or threat of violence) and the promise of salvation (or love). It was no good merely showing physical acceptance, the inner acquiescence of the victim was necessary, the mind had to give in. George Orwell in the last lines of *1984* described graphically such a final breakdown and acceptance: 'I love Big Brother'.

Napoleon's grip of France began in 1799 and the common people for the first time in European history had the sense of 'actively participating in the transformation of the world'. Military service became popular instead of a horror to be endured. The speculative thought of the intellectual was given the cold shoulder and talent was channelled into science and technology.

Such authoritarianism has been rife in the present century and all such systems have been suspicious of the speculative, critical thinker whose mind will not conform and who poses the awkward questions that make society restless and uneasy. These people have been terrorized, imprisoned and driven from their own country. Nevertheless their tormentors have always been sceptical about being able to change the victims' minds from an old value system to a new one. But not, apparently, the Chinese Communists. Their system of 'thought reform' makes, according to one authority 'a systematic assault on the inner identity of the victims, alternately bludgeoning and coaxing them into a mental confusion which blurs the lines between truth and falsehood, good and evil, inner self and external pressure, until in the end they realize that the only avenue of salvation is an acceptance of the thought-world of their tormentors'.

How long does the mind hold on to its beliefs as it is advanced to such a precipice of confusion? What are its gripholds as it clings on in the face of torture and deprivation? If, as Brock Chisholm writes, a study were made of episodes in history where the mind held on to its beliefs in the face of such desperate circumstances, we should certainly come to have a 'new faith in the indestructability of the human mind'.

30: Man's Behaviour

As we have seen man has an ordered memory of the past which gives him an expectation of the future and we have argued that the information stored in the brain through experience makes him think and act in certain ways. Man, we have seen, can talk about and pass on his learning to others. He can appreciate a beautiful sky, the smell of new-turned earth or a good painting, and he needs to worship God or idols. These are some of the characteristics which *appear* to separate him from the beasts of the field although he has plenty of behavioural characteristics which derive from an animal heritage. For example, his strong group attachments to a territory; his tendency to create closed societies with a definite peck-order; his aggressive behaviour to people behaving abnormally, and his own appeasement behaviour in the face of aggression.

Darwin, a century ago, was successful in showing how much of human behaviour has animal roots. Although the main line of development from lower animals to man is from a relatively rigid pattern of behaviour, constant within narrow limits, to a versatile and changing one, many animals are capable of much more than automatic behaviour. Some are 'terribly human'. Indeed Walt Whitman's inventory of animal 'virtues' no longer seems to square with observation. 'No mania for owning things' wrote Whitman, yet chimpanzees in captivity have been taught to work slot machines for a reward of grapes; then to obtain the grapes they had to insert counters. Two became money-conscious, hoarding piles of counters for future use. 'Not one kneels to another,' said Whitman, thinking of his servile fellow men. But the peck-order in hens is well-established and male baboons are known to bully and treat females brutally and even rape them; bribery of female chimpanzees in oestrus amounts to prostitution. 'Not one is unhappy', yet young goats display excessive signs of anxiety in stressful situations if they are not with their mothers, and an infant monkey behaves abnormally without its mother. Instead of exploring its surroundings confidently or sleeping peacefully it is very unhappy and rocks backwards and forwards in its anxiety and distress.

Even acquired experience can be passed on by imitation in monkeys, rats and chimpanzees, who, like humans, will not learn from just anyone. In the brown rat knowledge of a certain bait is passed on from generation to generation and the knowledge far outlives the rats who first contact the bait. Indeed according to Lorenz the difficulty of combating the brown rat, the most successful biological opponent to man, 'lies chiefly in the fact that the rat operates basically with the same methods as those of Man, by traditional transmission of experience and its dissemination within the close community'. Infant monkeys learn from their mothers rather than someone low in the pecking order. This learning from an admired or superior animal is familiar to us. If a schoolmaster, or anybody else for that matter, is not respected, he will not be able to teach successfully. And, as Professor Niko Tinbergen has pointed out, culture radiates from centres of power—large towns in mediaeval times, admired nations in modern times.

Communication by sound in animals is a familiar experience. Some of this is automatic and an appropriate stimulus arouses the response. Purposeful communication may have been observed in chimpanzees and the infant chimpanzee can be taught to utter two or three sounds which seem to have a specific meaning, e.g. 'cup'. In the human home chimpanzees can learn to receive and take messages by the use of symbols and gestures. But the gift of tongues and conceptual thought is unique to man. The chimpanzee, for all the tricks described above, has no built-in mechanism which enables it to translate the sounds that it hears into a complex mode of behaviour.

The upshot of all this shows first that there seems hardly any aspect of animal behaviour which does not throw some light on the problems of human behaviour, and second that the use of the word 'unique', suggesting that man is strikingly different from other animals, may be misleading. As Professor Tinbergen said—'on the one hand we recognize that Man is unique, but on the other hand we also recognize that he is an animal of a kind. One is greatly tempted to adapt an Orwellian observation and say no more than, "all animals are unique, but Man is more unique than others".'

The basis of this unique behaviour rests, as we have said, in man's large brain. But perhaps we should go a stage further than this to a lesson that cybernetics can teach us. It is as Professor N. Wiener says that the *structure* of the machine or of the organism is an index of the performance that may be expected from it. This means, for example, that the mechanical rigidity of an insect, imposed upon it by its hard outer skeleton, not only limits its size but also its mentality. Its size limits the number of nerve cells in its brain. It must be 'perfect' from the beginning of its adult life so that its brain must have the appropriate behaviour taped to see it through its future programme of life. The flexible structure and *physiology* of man, allows him novelty of experience and provides him with a brilliantly adaptable chassis and a motor to allow his brain almost indefinite expansion. As Wiener says, 'If we could build a machine whose mechanical structure duplicated human physiology, then we could have a machine whose intellectual capacities would duplicate those of human beings.'

Man's structure and physiology too, makes it necessary for him to spend a long and sheltered infancy learning from his own and others' experience. Civilized man is recognized as immature until he is twenty-one (over a quarter of his life) and often spends until he is 30 learning. So 40% of life is as a learner. Compare this with an insect which has to fly and feed and defend itself once out of its pupal case; or a cow that can run almost as soon as it is born.

METHOD AND LANGUAGE

To arrive at a better understanding of human nature, as Professor Niko Tinbergen has argued, we need to apply to man the methods and language of students of animal behaviour. The methods they use involve accurate descriptions of what animals *do* and what circumstances bring about these movements. The survival value of such movements (behaviour) is postulated and tested by experiment. In man, although movements can readily be described, the reasons for the movement (behaviour) are described in speech which might be grossly misleading. After all we might wish to cover up the real reasons for stealing something or striking somebody and thus the observable (speech) is worthless as a piece of evidence. Perhaps the subtleties of facial expression, of bodily postures and of slight inflexions of speech are the real equivalents of the data obtained from animals and are more reliable than words. As Professor Tinbergen points out, 'the fact that animals cannot tell us what urges them to behave as they do might be a blessing in disguise, at least they cannot tell us conscious or unconscious lies, they just behave'. Perhaps certain physiological data too, is more important than speech. The face may be serene and the tongue glib, yet the staggering or racing pulse, the rise in blood pressure betray the lie.

The language in which these data are described needs to be as objective as are the records of a student of animal behaviour. For example he may record that a male gull dashes at an opponent but stops before reaching him, then that he retreats and dashes again and withdraws, and so on. But if he records that the dash is in anger, then he is mixing up objective description with opinion which is subjective. No one knows whether the gull is angry or not. When human behaviour is described the objective, descriptive part is often mixed up with an interpretation of the observation. For example, 'he lit a cigarette nervously' is a simple mixture. For more complicated actions the difficult job is to sort out the description of observed movements from the subjective interpretation. If comparative studies are to be made on human and animal behaviour and if the comparisons are to be fruitful, then common methods and language are essential. By such comparative methods we should be able to learn more about our animal heritage and how it applies to man now. For example, we know little about aggression in man and its counterpart, fear. Once we understand these deeply-rooted urges we may be able to control them.

AGGRESSION

Studies of aggression in animals have shown that it is a deeply rooted instinct and that it serves a very useful purpose in the development of societies of mammals and birds. It and its counterpart fear cause animals of the same species to space themselves out and so prevent overcrowding during reproduction. Through aggression and fear the peck-order is developed in a closed society of animals, and as each learns its place fighting is reduced and individuals can spend their time profitably. When fighting does take place it is a civil war (that is between members of the same species) because of competition for food or mates. *Different* species of animals have their own food preferences and territories and keep to their own niche. This is like a human community, in a way for a butcher might compete with another butcher but not with a baker. Animals of the *same* species rarely harm or kill each other. Weaker animals retreat or run away from the stronger, or if cornered adopt a special 'appeasement posture'. A beaten wolf offers the nape of its neck to the eye-tooth of its opponent, a turkey cock lays out its neck for the beak of the victor to attack, a dog rolls on its back when defeated, a cock puts its head in a corner when beaten, thus removing from its opponent the fight-eliciting stimuli that the red comb and wattles stir up and curiously, these postures prevent the triumphant animal from hurting its rival. Instead of making the final kill, the aggressor may redivert his attack on to weaker brethren or even on to a lifeless object, a 'substitute' object. A dog may snap at the air and a black-headed gull may pull at the grass. Indeed species which have not developed these techniques have become extinct.

Man too seems to have his aggressive and appeasement postures. Lorenz notes that a railway carriage is an excellent place to observe the functions of aggression in the spacing out of territories in man. 'All the rude behaviour patterns serving for the repulsion of seat-competitors and intruders, such as covering empty places with coats or bags, putting up one's feet, or pretending to be asleep, are brought into action against the unknown individual only. As soon as the new-comer turns out to be even the merest acquaintance they disappear and are replaced by rather shamefaced politeness.' A beach too is another place to watch people staking territories—a beach to one's own family would be perfect!

The attitude when two men face each other with clenched fists and red faces resembles the aggressive upright posture of gulls (see Fig. 50); kicking a chair after a row or banging a fist on a committee table is a redirected attack on a 'substitute' object. Lighting a cigarette, doodling at a meeting, scratching the head, fiddling with keys are 'displacement activities' which are common in conflict situations in many animals including man. Perhaps, as Tinbergen writes, it might be worth explaining to what extent we can rechannel our aggression and sublimate it for instance by making concerted 'attacks' on nature, 'by conquering the sea, as the Dutch do, or by conquering space'.

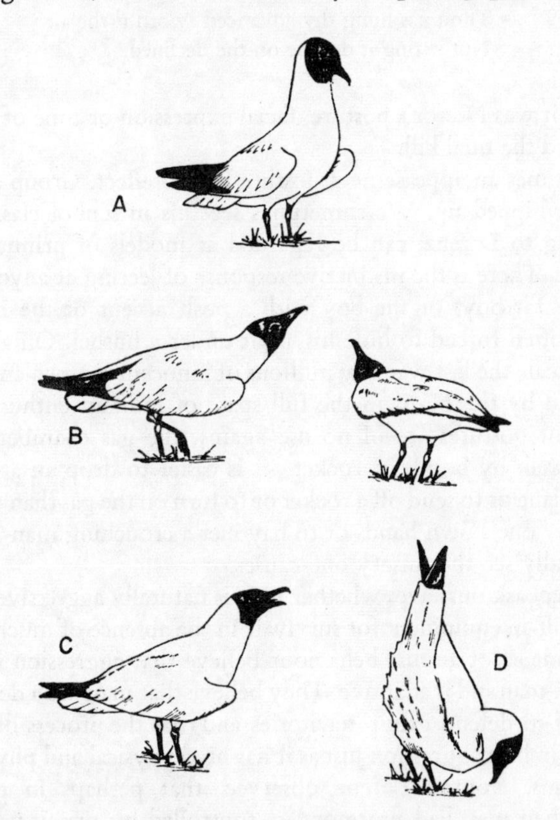

FIG. 50 Four agonistic postures of the black-headed gull. A, aggressive upright posture; B, forward posture; C, oblique posture; D, choking.

After Tinbergen, N., 'Aggression and fear in the normal sexual behaviour of some animals', in *The Pathology and Treatment of Sexual Deviations*, ed. I. Rosen (O.U.P., 1964).

Appeasement behaviour is common in man too. The baby's smile of the first few weeks and months is one behaviour pattern. Some mothers faced with an unwanted child soon change their attitude once smiling begins. And rarely do smiling men kill each other. Sometimes in war the trigger cannot be pulled because 'the inner obstruction' in the aggressor is unsurmountable. As Shakespeare makes Nestor say of Hector:

> Thou has hung thy advanced sword t' the air
> Not letting it decline on the declined.

Perhaps it was Nestor's posture, facial expression or tone of voice that prevented the final kill.

Sometimes an appeasement posture has no effect. Group aggression can be whipped up. We sometimes see this in school classes, which according to Lorenz, can be regarded as models of primitive group structure. There is the instinctive response of jeering at anyone abnormal: the fat boy, or the boy with a posh accent or the clever boy who is often forced to hide his light under a bushel. On a huge and terrible scale the last war saw millions of innocent women and children murdered by the Nazis in the full spate of militant enthusiasm. Appeasement postures are of no use against the gas chamber or long-range attack by bomb or rocket—it is easier to drop an atom bomb from a plane or to send off a rocket or to turn on the gas than to strangle a child by one's own hands or to bayonet a crouching man—one does not actually see the misery one causes.

We may ask ourselves whether man is naturally aggressive. Is it part of his built-in equipment for survival? In the absence of much evidence some students of animal behaviour believe that aggression is, indeed, inborn in man and is adaptive. They believe that man has a deep-rooted tendency to defend group territories and that the process of selection has bred in him aggression just as it has bred physical and physiological adaptations. Konrad Lorenz observed that perhaps in prehistoric times, when man had more or less controlled his primitive environment by weapons and tools, by clothing and fire, selection pressure shifted to produce a warrior breed. The necessity to produce them came from the need to fight off neighbouring hordes of men competing for food and territory. It was essential to avoid internal conflict within the community. Perhaps, as Lorenz states, the only commandment necessary to Cro-Magnon man was, 'Thou shalt not kill thy neighbour with a well-sharpened hand axe.' This was because it was necessary to save his strength to keep his territory intact in a life of war and raids. This kind of life still goes on between certain tribes of Papuans in Central New Guinea. In animals, the brown rat behaves similarly;

within each community rats behave tenderly towards each other, they are models of virtue, but stray rats and different packs of rats with different smells, are set on and murdered in the most bloody fashion.

There are peoples now who have in them these relics of aggression in excess. These are the Prairie Indians, the Utes whose aggressive traits were probably selected and fixed during the few centuries when they led a wild life consisting almost entirely of tribal war and later, war with the white settlers. The Ute Indians now live in reserves where there is little or no outlet for their aggression, and neuroses abound. They drive fast motor-cars and the rate of accidents exceeds that of any other car-driving human group. It is well known that accident proneness may result from pent-up aggression.

Those of us who live in 'civilized' countries may ask ourselves whether the cut-throat competition which goes on today between rival manufacturers of a similar product or whether such traits as 'thrustfullness, forcefulness', and self assertion, at a premium in our fast-moving society, may fix in us and our descendants characteristics as barbaric as those developed between warring sects of primitive man. If so then our society and those that evolve from it have a grim future—simple goodness will have no survival value but all those barbaric traits listed above will.

Right up to the present day stemming from the competition between warring tribes of Stone Age man, the power and prosperity of the family, sect, tribe or nation to which an individual belongs is of far more importance than the power, or even lives, of other individuals in other groups. And added to this dangerous faith is another: 'whenever we are frightened or feel threatened, the right, effective and virtuous thing to do is to increase our ability to kill other people'. These beliefs may seem primitive but now in an age of potential mass-destruction they still hold true and are doubly dangerous. For these and the other reasons stated above about our own competitive society, it is important to try to understand aggressive behaviour in man and how to control it. Why is it that we kill eagerly on a mass scale and to a plan? And on a more personal level why do we try, on occasion to edge another driver off the road? Why do we not like being overtaken by another car? (see Table 32.) Is aggressive behaviour essential to human progress? A beginning has been made with animals but as yet we do not know much about the signs that reveal in man the dagger of aggression let alone how to curb or control it.

The study of aggressive behaviour in man, therefore, might well

TABLE 32

Aggression in motorists

	Number	*Percentage*
Been in fight with another driver	28	7
Did not like being overtaken	53	14
Sometimes felt like killing another driver	53	14
Chased a driver who has annoyed	87	23

382 motorists were questioned in this investigation and some of the results are set out above. Data of Mr Meyer H. Parry, reported in *The Observer*, 15 January 1967. Mr Parry suggested that serious consideration needs to be given to personality characteristics in deciding whether a person is fit to drive.

bring its reward and indeed survival to our species in the long run, more so than by putting men into space or making a moon landing.

ANIMAL STUDIES

Research into aggressive behaviour in animals and man is fundamental and long-term. Many experiments have been made, and are being done now which are of direct value to the understanding of man's behaviour. It has been said before that we can learn something about ourselves from almost any aspect of animal behaviour but caution must be applied on any tendency to oversimplify. Comparisons can be grossly misleading especially if the simpler 'model' does not have the essentials of the more complicated system: a child's elastic-driven aeroplane would give one a very poor idea of how a real aircraft was powered. Even so animal studies can help in an essential task—to formulate hypotheses about human behaviour. Konrad Lorenz, a famous student of animal behaviour, has a special interest in fish and geese. He is interested in the way geese transmit information to each other by gesture and sound. The way for instance in which geese greet members of their own family ensures coherence of the family group and prevents attacks occurring within it. This has been learned by careful observation and tape recordings. As yet we do not know much about the way human beings differ in their gestures and tone of voice when they are with their own family, or among strangers. Such a study may help us to understand ourselves better and the geese studies may help to suggest fruitful paths of study (see Table 33).

Although we can learn a good deal about human behaviour through circumstantial evidence, human beings are poor material for any kind of experimental research. It is not possible to alter their genetic make-up by breeding experiments as we can in geese or fish or monkeys or try to bring about changes in behaviour by altering their environment. Nevertheless much can be learned about human behaviour from twin studies (see Chapter 5), by use of drugs as we have seen (Chapter 29), and by animal studies, especially with animals closely related to man in nervous organization like many mammals.

Animal studies with mice, rats, geese, rabbits and dogs have shown that not only nervous diseases are inherited but also differences in behaviour within the normal span of variation. Rats can be separated into those of high and low emotionality on the basis of the number of faecal pellets and urinary eliminations occurring under stressful situations such as variation in the amount of noise and light they are subjected to. There is, as we know only too well a direct relationship between fear and elimination and it has been shown that nervous and emotional parents breed nervous and emotional offspring. The young rats can be fostered with mothers of a non-emotional strain but the imprint of heredity on the fostered-rats over-rides any learned behaviour—the important factor is the genetic parentage.

Learning ability in rats, as in humans, is complex. It may depend on such genetically controlled variables as liveliness, exploratory activity and emotionality and many others. Much is heard by the general public about 'bright' and 'dull' strains of rats. The criterion here is the ability to learn to run through a maze in which alternate pathways lead either to a food supply or to a blind ending. It is better to call such rats 'maze-bright' or 'maze-dull', because (as in some I.Q. tests on human children) a maze-bright rat may obtain a low score in other learning tasks. One interesting experiment with maze-bright and maze-dull rats has meaningful implications for humans. The two strains were tested by rearing them in different conditions and then examining them for learning ability. The bright rats were not made brighter by rearing them in a stimulating environment; but dull rats improved in this enriched background and they reached the level of bright rats. Similarly dull rats brought up in restricted conditions were not made more stupid although bright rats so reared became dull rats.

TABLE 33

Some tentative comparisons between animal and human behaviour (based on the work of Professor N. Tinbergen[1] and Dr Konrad Lorenz[2])

Behaviour Type	Animal	Human
Territorial Behaviour Found most commonly in vertebrates. Adaptive: usually connected with reproduction. Spaces animals out and prevents overcrowding during reproduction.	Many examples. Mammals 'think through their noses' and leave scent marks at regular intervals wherever they go (e.g. cats). A cat finding another cat's signal on its hunting-path assesses its age, and if fresh it hesitates or goes another way. If it is a few hours old it carries on. Many animals will only fight in the middle of their own territory, their H.Q.	Passengers staking their claim in a railway compartment by placing handbags or papers on unoccupied seats; looking for an empty beach from which to bathe; fencing in a garden; 'keep out' notices; a feeling of confidence at home.
Rank, or Peck Order Found in higher vertebrates. Adaptive. Under this rule every individual in a closed society knows which one is stronger and which weaker than itself. The result is an enormous reduction of fighting, allowing individuals to spend their time profitably and saves the strength of the pack or group for attacks from *without.*	Many examples—e.g. Domestic fowl; Wolf pack; Monkey herd.	Tends to form closed societies with a definite peck order. Like all social animals man is a status seeker. Note the high tension between men who have *similar* positions in an office and note that tension diminishes the further apart the men are in rank.
Attacks on Abnormally Behaving Individuals Not well understood but probably adaptive and preserves the social norms characteristic of a group with are necessary for fighting, food gathering, reproduction and so on.	Many examples. In geese an individual that has recently flown into a colony may take years before it is accepted into a goose-group. Rat families hunt down and kill a strange rat.	Observed in schools: jeering at the fat boy, a clever boy, or a boy with a different accent. School classes and companies of soldiers 'models of primitive group structure'. The 'old school tie' effect serves to keep out an outsider or non-conformer. Absence of manners in an individual elicits anger and hostility in a group

Appeasement responses		
Found in higher vertebrates. Adaptive. In most animals appeasement gestures prevent aggression and therefore loss of life between members of *same* species.	Many examples. Wolf turns his head away from his opponent exposing to him his neck vein; Baboons present their behinds to a victorious opponent; Jackdaws hold the unprotected base of the skull under the beak of an aggressor; Macaques bare their teeth, smack their lips and lay back their ears.	Many forms of smile; certain tones of voice and postures; shaking hands, offering a cigarette; laughter; bowing; removal of hat; 'turning the other cheek'. Unfortunately appeasement gestures are of no use against push-button warfare, but only in hand to hand fighting.
Displacement Activities		
Activities which appear at first sight irrelevant in the situation in which they occur but always seen in conflict situations.	Many examples. Bill wiping, preening and feather movements in Black Headed gulls.	Various irrelevant movements such as fiddling with keys, scratching the head, smoothing the hair, yawning, smoking. All seen in conflict situations.
Attack on a Substitute Object		
The full strength of the attack is directed at 'taking it out' of an inanimate object or a weaker animal.	Many examples. Black Headed gulls peck violently into the ground; dogs worry a stick.	Seen in conflict situations. Banging the table with a fist or kicking a chair or shouting at the children after a bad day at work.
Rituals		
Serve to suppress fighting within the group, to hold the group together and to set it off as an independent entity against other similar units.	Many examples in insects and higher vertebrates. Courtship dances of many species of fly, triumph ceremony in geese.	Human behaviour full of rituals. Culturally developed social norms and rites are characteristic of human groups. Good manners which differ from culture to culture are strictly determined by cultural ritualization and are second nature. *Non-ritualized* behaviour is usually performed in private: nose picking, uninhibited yawning, scratching, would not be tolerated in public here. Children especially, cling to rituals. Like the savage they are respectors of custom. Note the ritual in street games and rhymes. Note that children quickly know when a favourite tale is not being told properly.

[1] Research Papers (see Bibliography). [2] *On Aggression*, Methuen, 1966.

MOTHERING

The two lines of evidence—circumstantial in humans and rigorous laboratory analysis of animal behaviour are enabling fruitful hypotheses to be made about human behaviour. These two lines are illustrated clearly by research on the working of that fundamental key-stone of society—the mother/child unit. Experiments with rhesus monkeys have shown how important this relationship is to the full expression of adult behaviour and in the development of normal sexual relations. Motherless monkeys were reared on substitute 'mothers' made of life-size, upright wire frames (like a tailor's dummy) some of which were covered with towelling. Milk was provided from a bottle attached to the frame. The babies preferred a soft, cuddly mother to a wire one even though the latter provided milk. They clung to the models, slept on them and in general used them as a safe base from which to explore and investigate their surroundings. Monkeys with no mother of any kind displayed anxiety and distress; they often rocked backwards and forwards huddled in a corner. But even so the young brought up on models did not learn to behave socially. Their sexual relationships were particularly disturbed. If one of them became pregnant its attitude to its baby was callous and indifferent. Males failed to copulate successfully. This behaviour is rather like that of severely disturbed people. Young children for example become severely disturbed if they are not attached to a single mother figure, and the evidence for this is abundantly clear in children who have lost their mother or who have had long stays in hospital or other institutions. D. J. Bowlby, observing a baby of six months to two years old and separated from its mother, in hospital or orphanage, notes three stages of behaviour. A period of protest; loud crying and expectations of the mother's return. This form of communication is of obvious biological survival value. If the separation is prolonged the child goes through the stages of 'despair' and 'denial'. A child in the third 'denial' stage can be won back to a normal relationship with his mother and other people only with great difficulty.

It is not only having a mother that is important but having one that allows the child to play with others. Experiments with rhesus monkeys show that what perhaps is missing in the model 'mothers' is the ability to reject the baby periodically so that it can learn, in the rough and tumble of play with its peers, proper social and sexual relationships. The model mothers seem to be 'over-protective'. While a good deal remains to be done in this field it is clear from this single example that work by experimental psychologists on animal behaviour will help to frame the questions we need to ask about our own behaviour.

It is appropriate that this book should end with a brief consideration of man's brain and behaviour. It is his brain that has created much of what we see around us. His mind, which is his brain, marks him off as human with aspirations, doubts, fears, longings, and responsibilities. His mind could either hurt his future as a species on earth or it could put him in a position where most men could live out of the shadow of hunger, poverty and fear. The choice is man's and is the final test of his abilities.

Bibliography

Some books and papers consulted in compiling this book. An asterisk indicates books which are suitable for the general reader.

PROLOGUE

Braun, Wernher von, 'Space Travel', *Weekend Telegraph*, 20 May 1966.
Dole, S. H., *Habitable Planets for Man*, Methuen, 1965.
Elton, Charles, *The Ecology of Invasions by Plants and Animals*, Methuen, 1958.
*Huxley, Julian, 'The future of man—evolutionary aspects', *Man and his Future*, C.I.B.A., Churchill, 1963.
Lorenz, K., 'On Aggression'. *Encounter*, August and September 1966.
*Thoday, J. M., 'Geneticism and environmentalism', *Biological Aspects of Social Problems*, Oliver & Boyd, 1965.

PART 1. HUMAN GENETICS. Chapters 1 to 10

Allison, A. C., 'Human haemoglobin types', *New Biology 21*, Pelican Books.
Auerbach, C., 'The genetic effects of radiation', *New Biology 20*, 1956.
Barnett, S. A., *The Human Species*, Pelican, 1957.
Beer, Gavin de, *Genetics and Prehistory*, The Rede Lecture, 1965. C.U.P., 1965.
*Carter, C. O., *Human Heredity*, Pelican, 1962.
Clarke, C. A., *Genetics for the Clinician*, Blackwell, 1962.

*Cooke, R. C., *Human Fertility*, Gollancz, 1951
*Darlington, C. D., *Genetics and Man*, Pelican, 1966.
—— 'Psychology, genetics and the process of history', *Brit. J. Psychol.*, 1963, **54**, 4.
—— 'Gypsies and Didikais', *Heredity*, 1959.
—— 'Cousin marriage and population structure', *Eugenics Review*, October, 1961.
—— 'Cousin marriage in man', *Heredity*, **14**, Parts 3 and 4, 1960.
—— 'Cousin marriages', *Triangle*, Vol. 3, No. 7, 1958.
—— 'Control of evolution in man, *Nature*, **182**, 1958.
—— *The genetics of society*, Private circulation, 1963.
—— 'The genetic component of language', *Heredity*, **1**, Part 3, 1947.
—— 'Speech, Language and Heredity', *Speech, Pathology and Therapy*, April 1961.
——'The chemistry of genetics', *M & B. Laboratory Bulletin*, February 1960.
——'Contending with evolution', *Science Progress*, January 1964.
Darlington, C. D. and Mather, K., *The Elements of Genetics*, Allen & Unwin, 1949.
Ferguson-Smith, M. A., 'Chromosome aberrations', *Biological Aspects of Social Problems*, Oliver & Boyd, 1965.
*Fisher, R. A., *The Genetical Theory of Natural Selection*, Dover, 1958.
*Galton, Francis, *Hereditary Genius*, Fontana, 1962.
*Haldane, J. B. S., *Heredity and Politics*, Allen & Unwin, 1938.
—— 'The prospects of eugenics', *New Biology*, 22.
—— 'Biological Possibilities in the Next Ten Thousand Years', *Man and his Future*, Churchill, 1963.
Harrison, G. A., *et al.*, *Human Biology*, O.U.P., 1964.
Huxley, Julian, 'The future of Man—evolutionary aspects', *Man and his Future*, Churchill, 1963.
—— 'The implications of modern genetics', *Listener*, 15 and 22 December 1949.
—— *Evolution—the Modern Synthesis*, Allen & Unwin, 1952.
—— *Evolution in Action*, Chatto & Windus, 1953.
*Medawar, P. B., *The Future of Man*, Methuen, 1960.
*Penrose, L. S., *Outline of Human Genetics*, Heinemann, 1963.
Sorsby, A., 'Noah—an albino', *B.M.J.*, December 1958.
Stern Curt, *Human Genetics*, Freeman, 1960.
Susser, M. W., and Watson, W., *Sociology in Medicine*, O.U.P., 1962.
Thoday, J. M., 'Geneticism and Environmentalism', *Biological Aspects of Social Problems*, Oliver & Boyd, 1965.
Today's Invalid Child—a brief survey, Invalid Children's Aid Association, 1966.

PART 2. RACE. Chapters 11 & 12

Baker, P. T., and Weiner, J. S., *The Biology of Human Adaptability*, O.U.P., 1966.
Coon, C. S., 'Growth and development of social groups', *Man and his Future*, Churchill, 1963.
★—— *The Living Races of Man*, Cape, 1966.
Haldane, J. B. S., *Heredity and Politics*, Allen & Unwin, 1938.
Howells, W., *Mankind in the Making*, Secker & Warburg, 1960.
Huxley, Julian, and Hadden, A. C., *We Europeans*, Cape, 1935.

PART 3. HEALTH. Chapters 13 to 19

Brock, J. F., 'Sophisticated diets and man's health', *Man and his Future*, Churchill, 1963.
★Burn, H. *Drugs, Medicine and Man*, Allen & Unwin, 1962.
★Burnet, M., *Natural History of Infectious Diseases*, C.U.P., 1953.
★Burnett, J., *Plenty and Want*, Nelson, 1966.
Card, Wilfrid, 'Towards a calculus in medicine', *Listener*, 9 February, 1967.
★Carstairs, G. M., *This Island Now*, Pelican, 1963.
★Comfort, A., *The Process of Ageing*, Weidenfeld & Nicolson, 1965.
★—— *Ageing—the Biology of Senescence*, Routledge, 1964.
—— 'Biology of Old Age', *New Biology*, 18. 1955.
Hudson, L., *Contrary imaginations*, Methuen, 1966.
—— 'Academic Sheep and Research Goats', *New Society*, 22 October 1966.
★Schmeck, H. M., *The Semi-Artificial Man*, Harrap, 1966.
Sherwood-Taylor, F., *Science Past and Present*, Mercury Books, 1962.
Susser, M. W. and Watson, W., *Sociology in Medicine* O.U.P., 1962.
Taylor, Lord, and Chave, S., *Mental Health & Environment*, Longmans, 1964.
★Walter, W. Grey, *The Living Brain*, Pelican, 1961.
Smoking and Health, Pitman, 1963.
'Georgian Centenarians', *Sunday Times Magazine*, 3 April 1966.

PART 4. POPULATION. Chapters 20 to 23

Burnet, M., *Natural History of Infectious Diseases*, C.U.P., 1953.
Calder, R., *Man and his Environment*, B.B.C. Publications.
Carson, Rachel, *Silent Spring*, Pelican, 1965.
Clark, Colin, 'Agricultural productivity in relation to population,' *Man and his Future*, Churchill, 1963.
Darlington, C. D., *The Genetics of Society*, Private circulation, 1963.
Fishlock, David, *Sunday Times Magazine*, 20 March 1966.

Florence, P. Sargant, 'Aspects of Fertility Control', *Biological Aspects of Social Problems*, Oliver & Boyd, 1965.
Harrison, G. A., *et al.*, *Human Biology*, O.U.P., 1965.
Hartley, Sir Harold, *Nutrition: The World's Problem*, Seale-Hayne Agricultural College, Newton Abbot, Devon.
Hutchinson, J. B., and Wilman D., 'The Strategy of Food Production', *Discovery*, May 1965.
Huxley, Julian, 'The future of man-evolutionary aspects', *Man and his Future*, Churchill, 1963.
Langer, William L., 'The Black Death', *Scientific American*, February 1964.
Moore, N. W., Toxic Chemicals and birds, *British Birds*, 55, 1962.
—— 'Environmental contamination by pesticides, *Ecology and the Industrial Society*, Blackwell, 1965.
Morgan, Dorothy, 'Birth Control Service', *Biological Aspects of Social Problems*, Oliver & Boyd, 1965.
*Mumford, Lewis, *The City in History*, Secker & Warburg, 1961.
Parkes, A. S., 'The future of fertility control', *Biological Aspects of Social Problems*, Oliver & Boyd, 1965.
Wiener, N., *The Human use of Human Beings*, Eyre & Spottiswoode, 1955.
Traffic in Towns, H.M.S.O., 1963

PART 5. YOUTH AND AGE. Chapters 24 to 27

Casey, M. D., *et al.*, 'Chromosomes and crime,' *World Medicine*, Vol. 2, No. 11, 1967.
*Comfort, A., *Sex and Society*, Pelican, 1964.
*—— *Ageing—the Biology of Senescence*, Routledge, 1964.
*Darlington, C. D., *Genetics and Man*, Pelican, 1966.
*Eysenck, H. J., *Crime and Personality*, Routledge, 1964.
Griffiths, Eldon, M.P., 'Armchairs versus Motorbikes', *Daily Telegraph*, June, 1966.
Lambert, Royston, *Nutrition in Britain*, 1950–60, Codicote Press, 1964.
Lange, J., *Crime as Destiny*. Allen & Unwin, 1931.
Schofield, Michael, *The Sexual Behaviour of Young People*, Longmans, 1965.
Stengel, E., 'Facts, Figures and Suicide', *Discovery*, 1964.
Tanner, J. H., *Human Biology*, ed. Harrison, *et al.*, O.U.P., 1965.
Teenage Morals, Education Press, London.
Thrasher, F. M., *The Gang*, Univ. Chicago, 1927.
Townsend, Peter, 'Old Age', *Observer*, 18 July 1965.
*Trotter, W., *Instincts of the Herd in Peace and War*, Allen & Unwin, 1965.
Wootton, Barbara, *Social Science and Social Pathology*, Allen & Unwin, 1959.

PART 6. BEHAVIOUR. Chapters 28 to 30

Barnett, S. A., 'The behaviour and needs of infants mammals', *Lancet*, 20 May 1961.

Broadhurst, P. L., 'The inheritance of behaviour', *Science Journal*, June 1965.

—— 'Animal behaviour and mental health', *New Society*, 9 July 1964.

*Cannon, W. B., *The Wisdom of the Body*, London, 1932.

Chisholm, B., *Can People learn to learn*, Harper, 1958.

Cullen, M., 'Behaviour', *Science in its Context*, Heinemann, 1964.

Fuller, J. L., and Thompson, W. K., *Behaviour Genetics*, Wiley, 1960.

Eccles, Sir John, *The Brain and the Unity of Conscious Experience*, C.U.P., 1965.

Farber, S. M., and Wilson, K. H. C., ed. *Control of the Mind*, McGraw-Hill, 1961.

*Huxley, A., *The Doors of Perception*, Chatto and Windus, 1959.

*Lorenz, K., *King Solomon's Ring*, Reprint Society, 1953, also *Encounter*, August and September 1966.

—— 'Man and the Sociable Goose in *Observer's Weekend Review*, 1 May, 1966.

*—— *On Aggression*, Methuen, 1966.

Mace, C. A., 'The mind and society', *The Control of the Mind*, ed. Farber, McGraw Hill, 1961.

McKay, D. M., 'Cybernetics', *Science in its Context*, Heinemann, 1964.

Sherrington, C. S., *Man on his Nature*, C.U.P., 1951.

Tinbergen, N., 'The Search for animal roots of human behaviour', Lecture given in a series *Social Studies and Biology*, 27 October 1964.

—— 'Aggression and fear in the normal sexual behaviour of some animals', *The Pathology and Treatment of Sexual Deviation*, ed. I. Rosen, O.U.P., 1954.

*Walter, W. Grey, *The Living Brain*, Pelican, 1961.

Weiner, N. *The Human Use of Human Beings*, Eyre & Spottiswoode, 1954.

Young, J. Z., *A Model of the Brain*, Oxford, 1965.

*—— *Doubt and Certainty in Science*, Oxford, 1951.

Index

A.I.D. 78
Adopted children 44, 52–3
Adrenaline, and mescaline 235
Ageing 143
Aggression 241
  animal 246
  motorists 245–6
Aggressive tendencies xviii
Agriculture, improved 160–1
Albinism 9, 59
Alkaptonuria 11, 22
Appeasement 242
Aptitude tests 148
Art, children's 228
Assortative mating 51, 56

Bach, pedigree 39
Bedford, Mass Survey 109, 117, 119
Behaviour, animal and human 238
Biological engineering 93
Birds, injury to 179–80
Birth, weight at 40
Birth control 169, 171
Blood groups 85–6
  and disease 68, 70
  of Europe 88–9
  and language 87
  Welsh 86–7
Body build, and criminality 205–6
  and disease 121
Brain, metabolism 225
  model of 227
Bushmen 94
Butterflies, decline of 180

Cancer, cervical 119–20
  National differences 120
Chimpanzee behaviour 239
Chlorpromazine 115–16
Cholesterol 105
Chromosomes 3, 5
  breakages 34
  criminal behaviour and 201
  extra 26, 47
  grafting proposed 78
Cities, growth 186–7
  origins 191
Class structure, and marriage 56
Cleft palate 71, 77
Climate, and race 92
Colour blindness 29, 30
Comfort, Alex 140, 142–3, 145, 218
Computers 224
Continuous characters 17
Coronary thrombosis 103
Cousins, marriage of 9, 57
Crime, incidence of 197–8
Criminals, genetics 199, 200
Crops, origin 73
Curiosity 230
Cybernetics 240
Cystic fibrosis 76

D.M.P.E. 235
D.N.A. 21
Darwin, Charles 64

Darwin-Wedgwood-Galton pedigree 43, 45
Deaf-mutism 11
Death rates, male and female 27–8, 32
Diabetes 109
  mass survey 119
  water 29
Didikai 97
Diet, labourers' 138
Diphtheria 110
Disease, changing patterns 109
  and population 153
  of undeveloped areas 123
  unrecognized 117
Disruptive selection 73
Dominant factors 5, 7, 11
Drosophila, breeding in 17
Drugs, of escape 233

EGGS, poisoning of 180
Environment and genius 51
Environmental effect 19, 38
Enzymes 21, 23
Eskimo 93
  diet of 131–2
Eugenics, negative 77
  positive 78
Evolution 65
  human 66
Eye colour 6, 7

FAMILY EXTINCTION 61
Fertility, control of 167–8
  and death rates 158–9
Food, additives 135–6
  adulteration 137–8
  new sources 162
  preservation 136

GALACTOSAEMIA 24
Galton, Francis 17, 40, 42, 49, 91
Gametes 4
Garrod, Archibald 22
Genes 5
  autosome 26
  Duchenne type 29
  plus and minus 16
  sex-linked 26, 28
  shuffling of 13
  size of 21
  working of 20
Genetic group difference 66
Genius 49
Georgetown, D.D.T. treatment 159
Georgia, life span in 145
German race 85
Gout, dominance 12
Group behaviour 203
Gypsies 45, 97
  Hungarian 87

HABITAT, conservation xix
Haemoglobin, and malaria 68
Haemophilia 29, 30
  mutant rate 36
Haldane, J. B. S. xviii, 56, 78, 98, 101, 115, 120, 148
Hallucinogens 234
Handicaps, after infectious disease 72
Hapsburgs, lip of 11
Hare lip 71, 77
Hedgerows 179
Height, human 17
Heiress marriages 61–2
Hiroshima 34–5
Hottentots 94
  palate shape 90
Huntington's chorea 13
Huxley, Julian xvii, 64, 66
Hybrid vigour 143
Hysteria, in dogs 138

I.Q. xviii, 44, 52
  tests 90
India, diets of 132
Individuality, human xviii, 38
Inequality of man 17
Inheritance of characters 5
Insecticides 183
Insulin 70, 109
Intelligence and race 90
Intercourse, reasons against 213
Irish famine 159

JEWS 84
Judges, inheritance 50

LAND USE 174–7
Language, and blood groups 87
Lederberg, Joshua 78
Life, average span 141, 146
  in Georgia 145
Limbic system 235
'Lobster claw' deformity 13
Lorenz, Konrad xviii, 239, 242, 244, 246, 248
Lung cancer 107–8

MALARIA, genetic protection 95
  sickle cell anaemia 67
Mass screening 118, 120
Meals, 'perfect' 131
'Megalopolis' 193
Mellanby, Edward 138
Memory 227
Menarche, year of 209, 211
Mendel, Gregor 3
Mendel's laws 5, 9
Mental deficiency, genetic 24
Mental disease 112
Metalworking, beginning of 73
Mongolism 33, 44, 47–8
Moths, selection in 65
Mothering 250
Motor cars, numbers 189
  traffic xviii
Muscular dystrophy 30
Mutations 23, 33
  artificial 34
  and evolution 37
  spontaneous 34

NAPOLEON 237
Natural selection 64–5
'Newtown', neurosis in 112, 114
Noah 9, 10
Nucleic acids 21

OBEDIENT CHILDREN 236–7
Obesity 136
Old, care of xix
  problems of 215–16
Orwell, George 237, 240
Overcrowding, stress from 191–2
Overpopulation xvi

PACEMAKER, of brain 231–2
Pancreas, fibrocystic disease 11
Peck order 192, 239, 247
Perception 223
Personality, and disease 121–2
Phenylketonuria 11, 24, 59
Phosphenes 228–9
Pills, contraceptive 168
Plague 154, 187
Planets, habitable xv
Plastic coil 170
Poliomyelitis 110–11
Polygenes 15*f*, 16, 17
Population, checks 154
  control of xvii
  fast growing areas 166
  growth 151
  problems xvii
Porcupine defect 12, 29
Pregnancy, length of 17
Problem families 173
Promiscuity 212, 213
Proteins 21
  deficiency 162
  dietary 125
  future requirements 166
  hunger 132, 133
  leaf 162
Pygmies 94
Pyloric stenosis 70, 71

R.N.A. 21
  and memory 230
Race, crossing 96
  and intelligence 90
  meaning 98
Rat, behaviour 245
  hereditary memory 239
Recessive factors 5, 7, 9
Recombination 13, 14
Rickets 133
River pollution 178

Schizophrenia, pink spot 235
Scipio family 11
Seas, as food source 163
Segregation of genes 5
Sex determination 25

Sex-linked diseases 31
Sex ratio 27
Sickle cell 66
  haemoglobin 23
Skipping a generation 7
Smoking, and disease 108
Social evolution 73
Struldbruggs 220
Sugar, in diet 105
Suicide 113
  of elderly 216
Surgery, spare-part 146

TH SOUND 87–8
Thalassemia 67
Time genes 13
Tinbergen, Niko 239, 240–1
Tissue grafts 146
Tranquilizers 114–15
Translocation of chromosomes 48
Tropics, agriculture 160
Twins 40, 52
  criminality in 199, 200
Typhoid 156–7

UTE Indians 245

V.D. 208
  in young people 213
Victoria (Queen), haemophilia in 30, 33
Voice prints 18

WATER, desalinification 164–5
  supplies 164
Wild life, loss of 181–2, 184–5
William the Conqueror 58
'Witchcraft', genetic 13

X CHROMOSOMES 25
X-rays, mutation effects 35

Y CHROMOSOMES 26
Yeast as protein source 162
Youth, extension of 140
Yudkin, John 103, 105

ZULUS 127